Collapse Calderas Workshop 2008

IOP Conference Series: Earth and Environmental Science
Volume 3

Querétaro, Mexico
19 - 25 October 2008

ISBN: 978-1-5108-0207-0
ISSN: 1755-1307

Printed from e-media with permission by:

Curran Associates, Inc.
57 Morehouse Lane
Red Hook, NY 12571

Printed by Curran Associates, Inc. (2015)

For permission requests, please contact the Institute of Physics
at the address below.

Institute of Physics
Dirac House, Temple Back
Bristol BS1 6BE UK

Phone: 44 1 17 929 7481
Fax: 44 1 17 920 0979

techtracking@iop.org

Additional copies of this publication are available from:

Curran Associates, Inc.
57 Morehouse Lane
Red Hook, NY 12571 USA
Phone: 845-758-0400
Fax: 845-758-2634
Email: curran@proceedings.com
Web: www.proceedings.com

Table of contents

Volume 3

COLLAPSE CALDERAS WORKSHOP
19–25 October 2008, Querétaro, Mexico

PAPERS

Collapse Calderas Workshop IOP Publishing
IOP Conf. Series: Earth and Environmental Science **3** (2008) 012001 doi:10.1088/1755-1307/3/1/012001

Interpreting compositional zonation of the Zaragoza ignimbrite from Los Humeros caldera, central Mexico

Carrasco-Núñez[1], Gerardo, McCurry, Michael[2], Branney, Michael J[3]

[1] Centro de Geociencias, UNAM, Campus Juriquilla, Querétaro, Qro., México.
[2] Department of Geology, Idaho State University, Pocatello, ID, USA
[3] Department of Geology, University of Leicester, Leicester, UK

Keywords: compositional zoning, ignimbrite, pumice, mixing

Abstract

Compositional zonation in ignimbrites is relatively common, and is often inferred to record gradual withdrawal by an eruption of a density-stratified magma chamber (with silicic magma towards the top and more dense, mafic magma at the bottom). This simple model was initially proposed to explain the complex chemical zonation of the (c. 0.1 Ma; DRE vol 15 km^3) Zaragoza ignimbrite from Los Humeros caldera in central Mexico. However, detailed petrologic studies reveal a more complex scenario: the ignimbrite exhibits a 'double' vertical zonation based on the compositions of pumice lapilli: a basal zone dominated by rhyodacitic pumice (68 wt.% SiO_2) changes gradationally up into a central zone with abundant andesitic pumice (54-62 wt.% SiO_2) zone, and this passes gradationally up into an upper rhyodacitic zone that lacks andesitic pumice. Petrographic and microprobe analyses of coexisting glass and phenocrysts provide evidence of equilibrium and disequilibrium conditions in the magmas at the time of eruption. The andesitic pumices contain euhedral labradorite (~An_{60}), and ortho- and clinopyroxene, in a dacitic glass matrix, that yield equilibrium ${}^{Na\text{-}Ca}K_d^{pl/liq}$ and ${}^{Fe\text{-}Mg}K_d^{px/liq}$ ratios. However, they also contain highly calcic plagiocase (to An_{82}) that in some cases is highly resorbed and mantled by the more sodic plagioclase, and this disequilibrium assemblage may have been produced by pre-eruptive mixing between andesitic and plagioclase-bearing basaltic magmas, followed by equilibrium crystallization within the hybrid magma. The rhyodacitic pumice contains more evolved plagioclase (~$An_{30\text{-}40}$) and pyroxene in a rhyolitic glass matrix (74-75% SiO_2). The plagioclase and pyroxene are both are euhedral, but the latter yield disequilibrium ${}^{Fe\text{-}Mg}K_d^{px/liq}$ ratios which indicate that they formed from a liquid that was more mafic that the pumice glass. Many pumices exhibit intricate interbanding of rhyodacitic and andesitic end-members. That, and the paucity of intermediate glass, and bulk pumice compositions suggests that that mingling and limited mixing occurred during or immediately before the caldera-forming eruption. One possibility is that the ignimbrite eruption occurred in response to intrusion of a hybridized andesitic magma into a rhyodacitic magma chamber.

IOP Conf. Series: Earth and Environmental Science **3** (2008) 012002 doi:10.1088/1755-1307/3/1/012002

New geologic evidence for additional 16.5-15.5 Ma silicic calderas in northwest Nevada related to initial impingement of the Yellowstone hot spot

Matthew A. Coble and Gail A. Mahood
Dept. Geological and Environmental Sciences, 450 Serra Mall, Bldg 320, Stanford University, Stanford, CA 94305-3115, USA

Keywords: Northwest Nevada, Calderas, Peralkaline, Ignimbrite, Yellowstone

1. Introduction

Although McDermitt caldera, on the Nevada-Oregon border (Figure 1), is frequently shown as the site of initial impingement of the Yellowstone hot spot, new mapping demonstrates that the area affected by this mid-Miocene volcanism is significantly larger than previously appreciated.

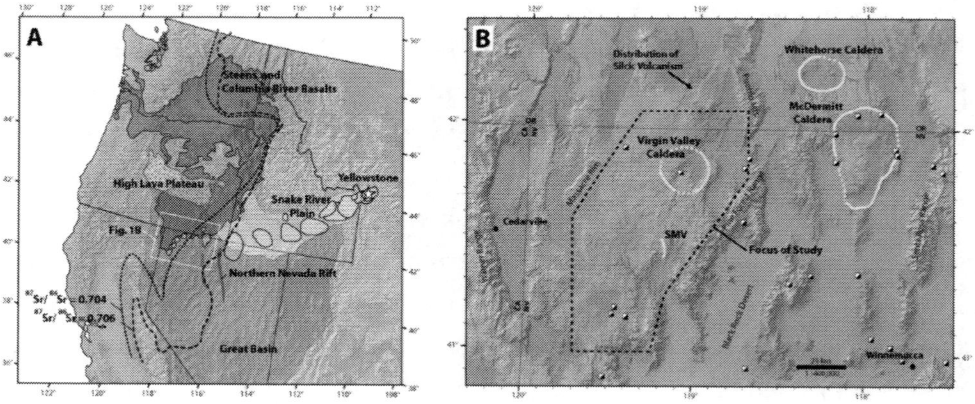

Figure 1. (A) Location map for silicic centers related to the Yellowstone hot spot in the western United States. The calderas (yellow) lie near the southern end of the distribution of the Pueblo, Steens, and Columbia River flood basalts (brown), and mark the intersection of basaltic dikes (red) of the same age and the geophysical anomaly of the Northern Nevada Rift zone. Silicic volcanism youngs to the east in the Snake River Plain. In the High Lava Plateau, the inception of silicic volcanism youngs westward. From Rytuba and McKee (1984), Zoback *et al.* (1994), Bussey (1996), Camp and Ross (2004), Jordan *et al.* (2004), Coe *et al.* (2005), Nash *et al.* (2006), and Brueseke *et al.* (2007). (B) Location map showing mineral deposits (boxes), the known extent of mid-Miocene rhyolitic ignimbrites and lavas (green shading), and the location of mid-Miocene calderas: Whitehorse and McDermitt from Rytuba and McKee (1984), Virgin

Valley from Castor and Henry (2000) outlined in yellow; newly identified by our mapping outlined in orange. The Soldier Meadows lava vent alignment (SMV) of Korringa (1973) is also shown. Mineral deposits from Willden (1964), Rytuba *et al.* (1979), Castor and Henry (2000), and Wallace *et al.* (2004).

2. Analysis

Three silicic calderas have been newly identified in northwest Nevada west of McDermitt caldera (Fig. 1b). Comparing the stratigraphic relations of caldera-forming units to the $^{40}Ar/^{39}Ar$ ages of Castor and Henry (2000), these calderas, along with the Virgin Valley, Whitehorse and McDermitt calderas (Rytuba and McKee, 1984; Castor and Henry, 2000), are interpreted to have formed during a short interval at 16.5-15.5 Ma, during the waning stage of Steens flood basalt volcanism. Outcrops of high-temperature, peralkaline rhyolite ignimbrites occur over approximately 5,000 km^2. Vents for high-temperature, low-viscosity, post-caldera peralkaline rhyolite lavas are interpreted to delineate the location of the now-buried caldera ring faults. The calderas have diameters ranging from 15 to 26 km.

The calderas lie along a NNE trend near, and parallel to, the inferred edge of the North American craton. The calderas occur at the southern end of the distribution of feeder dikes for the Columbia River and Steens basalts, suggesting a relationship between the location of dike-fed flood basaltic volcanism and the development of caldera magmatism in the mid-Miocene. Economically important uranium, lithium, and epithermal gold deposits are associated with these mid-Miocene calderas (Rytuba *et al.*, 1979; Rytuba, 1989; Bussey, 1996; Castor and Henry, 2000; John 2001), occurring in lacustrine sediments and ring-fracture rhyolite lavas (Figure 1b).

Fine-grained caldera lake sediments are preserved in all of these mid-Miocene calderas, indicating how little eroded the calderas are. We have identified major canyons cut by the overtopping of the caldera lakes, suggesting that the region was elevated during the mid-Miocene. The block of crust encompassing the calderas has undergone much less Basin-and-Range normal faulting than regions to the east and west, suggesting that underlying plutons may have made the upper crust in this region stronger than in adjacent areas.

REFERENCES

Brueseke, M. E., Heizler, M. T., Hart, W. K., and Mertzman, S. A., 2007, Distribution and geochronology of Oregon Plateau (U.S.A.) flood basalt volcanism: The Steens Basalt revisited: Journal of Volcanology and Geothermal Research, 161, 187-214.

Bussey, S. D., 1996, Gold mineralization and associated rhyolitic volcanism at the Hog Ranch District, northwest Nevada, *in* Coyner, A. R., and Fahey, P. L., eds., Geology and Ore Deposits of the American Cordillera: Geological Society of Nevada Symposium Proceedings: Reno/Sparks, Nevada, April, 1995.

Camp, V. E., and Ross, E. R., 2004, Mantle dynamics and genesis of mafic magmatism in the intermontane Pacific Northwest: Journal of Geophysical Research, 109, doi: 10.1029/2003JB002838.

Castor, S. B., and Henry, C. D., 2000, Geology, geochemistry, and origin of volcanic rock-hosted uranium deposits in northwest Nevada and southeastern Oregon, USA: Ore Geology Review, 16, 1-40.

Coe, R. S., Stock, G. M., Lyons, J. J., Beitler, B., and Bowen, G. J., 2005, Yellowstone hotspot volcanism in California? A paleomagnetic test of the Lovejoy flood basalt hypothesis: Geology, 33(9), 697-700.

John, D. A., 2001, Miocene and Early Pliocene epithermal gold-silver deposits in the northern Great Basin, western United States: characteristics, distribution, and relationship to magmatism: Economic Geology, 96, 1827-1853.

Jordan, B. T., Grunder, A. L., Duncan, R. A., and Dcino, A. L., 2004, Geochronology of age progressive volcanism of the Oregon High Lava Plains: Implications for the plume interpretation of Yellowstone: Journal of Geophysical Research, 109, doi: 10.1029/2003JB002776.

Korringa, M. K., 1973, Linear vent area of the Soldier Meadow Tuff, an ash-flow sheet in northwestern Nevada: Geological Society of America Bulletin, 84, 3849-3866.

Nash, B. P., Perkins, M. E., Christiansen, J. N., Lee, D.-C., and Halliday, A. N., 2006, The Yellowstone hotspot in space and time: Nd and Hf isotopes in silicic magmas: Earth and Planetary Science Letters, 247, 143-156.

Rytuba, J. J., 1989, Volcanism, extensional tectonics, and epithermal mineralization in the northern Basin and Range province, California, Nevada, Oregon, and Idaho, *in* Schindler, K. S., ed., USGS research on mineral resources - program and abstracts: Fifth Annual V.E. McKelvey Forum on Mineral and Energy Resources., U.S. Geological Survey Circular, 59-61.

Rytuba, J. J., Glanzman, R. K., and Conrad, W. K., 1979, Uranium, thorium, and mercury distribution through the evolution of the McDermitt caldera complex., *in* G.

W. Newman, e., Rocky Mtn. Assoc. Gaol. Guidebook., ed., Basin and Range symposium and Great Basin field conference, 405-412.

Rytuba, J. J., and McKee, E. H., 1984, Peralkaline ash flow tuffs and calderas of the McDermitt Volcanic Field, southwest Oregon and north central Nevada: Journal of Geophysical Research, 89(B10), 8616-8628.

Willden, R., 1964, Geology and mineral deposits of Humboldt County, Nevada: Nevada Bureau of Mines and Geology Bulletin, 59, 1-154.

Zoback, M. L., McKee, E. H., Blakely, R. J., and Thompson, R. A., 1994, The northern Nevada rift: regional tectono-magnetic relations and middle Miocene stress direction: Geological Society of America Bulletin, 106, 371-382.

Pucarilla-Cerro Tipillas Volcanic Complex: The oldest recognized caldera in the southeastern portion of Central Volcanic Zone of Central Andes?

Silvina Guzmán[1*] and Ivan Petrinovic[1]
[1] CONICET –IBIGEO. Museo de Cs. Naturales, Universidad de Salta, Mendoza 2 (4400), Salta, Argentina. *guzmansilvina@gmail.com

Keywords: Pucarilla-Cerro Tipillas Volcanic Complex, CVZ, Andes, Miocene Magmatism, Caldera.

1. Introduction

In the southeastern portion of Central Volcanic Zone of the Central Andes there are vast outcrops of pyroclastic and minor effusive deposits which lack of previous volcanological and petrological studies. We group individual pyroclastic deposits into the Pucarilla-Cerro Tipillas Volcanic Complex (PCTVC), relating them to a caldera structure.

Deposits from this complex are located in the southeastern margin of the Puna Plateau, constituting one of the first important expressions of magmatism in this zone. This might be important because it coincides with the onset of major uplift in this region of the Puna (Isacks 1988; Allmendinger *et al.* 1997), thus contributing to crustal thickening which is in part attributed to magma addition (e.g. Allmendinger *et al.* 1997).

2. Results

Units of PCTVC lay unconformably above Paleozoic granitic basement and folded sediments (Marrett *et al.* 1994; Hongn and Seggiaro 2001) of the Eocene-Pliocene Payogastilla Group.

We include and group into the PCTVC the deposits that were alternatively assigned to *Laguna Blanca Formation* (Turner, 1962), "*Tobas dacíticas*" (Navarro García, unpublished), "hornblendic welded tuff" (Sparks *et al.*, 1985), "*flujos ignimbríticos*" (Castillo and Aguilera, 1988), "*andesitas y dacitas Tebenquicho*" (González, 1992), "welded tuff" (Marrett *et al.*, 1994), "*Ignimbrita Pucarilla*" (Hongn and Seggiaro, 2001) and "*Ignimbrita Pampallana*" (González *et al.*, 1999).

Most of the recognized deposits are located in a NNE-SSW elongated inter-mountain relief

IOP Conf. Series: Earth and Environmental Science 3 (2008) 012003 doi:10.1088/1755-1307/3/1/012003

(25°49'-26°37'S and 66°16'-66°43'W). Three massive pyroclastic lithofacies are recognized as medium to high grade ignimbrites, the most developed of which is *Pucarilla* Ignimbrite. The latter corresponds to the most welded and devitrified pyroclastic unit and was dated at 12.11 +/- 0.11 Ma (Marrett *et al.*, 1994)

Some authors as Hongn and Seggiaro (2001) and González *et al.* (1999) have suggested a relationship between *Pucarilla* Ignimbrite and the hornblendic welded tuff dated in 15.83 +/- 0.44 Ma and 14.75 +/- 0.4 Ma by Sparks et al. (1985). Cropping out in the surroundings of Alto de Las Lagunas depression, this unit is a pink-welded and devitrified ignimbrite that needs to be considered as a separated unit, due to slight differences in its petrography. Nevertheless, it is texturally and compositionally very similar to *Pucarilla* Ignimbrite; for this reason we include it in the PCTVC as a single unit. The stratigraphic relation between both units is still unknown.

Compositionally, all the recognized units are metaluminous to peraluminous andesites and dacites of the calk-alkaline series and are characterized by a high crystal (33-48%) and low lithic content (<3%); normal pumices/fiammes content varies between 5 and 35%. As a characteristic feature all of them have pink fiammes (< 1%) and grey juvenile fragments (<1%). We did not recognize all outcrops possibly related to the complex, so areas farther west and south of the interpreted emission centre should be examined and might increase notably the areal distribution and volume of the PCTVC.

3. Characterisation

Major characteristics of the PCTVC are: (i) most deposits are intensely welded tuffs (ii) their welded characteristics persist at least for distances of 50 km, (iii) no initial plinian deposits are recognized, (iv) the magmatic event ends with effusive activity, (v) extruded magma volume is at least 20 km^3, (vi) the minimum areal distribution is 650 km^2 (vii) the greatest actual topographic difference between the deposits of the complex is about 2,2 km. In light of the described characteristics and a detailed field work, discriminating and correlating facies, we invoke a volcanic structure such as a caldera being responsible for this succession.

We propose a probable centre linked to the depression of Alto de Las Lagunas (26° 09'S-66° 43'W) where some field evidences can be related to a non resurgent collapse caldera. These include: (i) an intense hydrothermalization in the borders of a topographic depression which is (ii) surrounded by pyroclastic units of PCTVC and (iii) lavas in proximal facies, (iv) absence of basement (paleozoic granites).

Deposits from PCTVC have an age averaging between 15 and 12 Ma, revealing that the magmatic activity in this region is more developed than previously thought. This caldera becomes the easternmost and one of the oldest known examples in the Central Volcanic Zone of Central Andes.

4. Conclusions

Given the spatial proximity between Galán Caldera and PCTVC it should be interesting to evaluate their relationship, which is still unclear given the few complete geochemical published data. Nevertheless a first approximation shows: a) bivariate diagrams show possible correlatable trends, b) rare earth elements patterns show more enrichment in LREE and depletion in HREE in Galán rocks compared to PCTVC c) multi-element patterns are rather similar. For this, although they have somewhat different geochemical patterns, they could indicate a recurrence in the volcanism between 12 and 2 My in the south eastern end of the Altiplano Puna plateau. The magma signature of both eruptive periods (12 and 2 Ma), could be useful to test the geochemistry as a tool, in order to gain insight into the knowledge of crustal thickening. Detailed comparisons of them are still under study.

Considerable work remains to be done to elucidate the history of this volcanic complex and its relationship with the evolution of the southeastern portion of the Altiplano-Puna plateau.

REFERENCES

Allmendinger, R. W., Jordan, T.E., Kay, S. M., Isacks, B. L., 1997, The evolution of the Altiplano-Puna Plateau of the Central Andes: Annual Review of Earth and Planetary Sciences (25), 139-174.

Castillo, A.L., Aguilera, N., 1988, Geología Económica de Jasimaná, Provincia de Salta in III Congreso de Geología Económica, Olavaria, Argentina, 347-367.

González, O. E., 1992. Geología de a Puna Austral entre los 25° 15'a 26° 30' de latitud Sur y los 66° 25'a 68° 00' de Longitud Oeste, Provincias de Catamarca y Salta, Argentina: Acta Geológica Lilloana XVII (2), 63-88.

González, O. E., Viruel, M. E., Fernández, D. S., 1999. El Complejo Piroclástico Cerro Tipillas en el borde oriental de la Puna Austral, Argentina in XIV Congreso Geológico Argentino, Actas II, Salta, 236-239.

Hongn, F., Seggiaro, R., 2001. Hoja Geológica 2566-III Cachi, provincias de Salta y Catamarca. Instituto de Geología y Recursos Minerales, Servicio Geológico Minero Argentino, Boletín 248, Buenos Aires, 87 p.

Isacks, B., 1988, Uplift of the Central Andes plateau and bending of the Bolivian Orocline: Journal of Geophysical Research, (93), 3211-3231.

Marrett, R., Allmendinger, R.W., Alonso, A.N., Drake, R.E., 1994, Late Cenozoic tectonic evolution of the Puna plateau and adjacent foreland, northwestern Argentine Andes: Journal of South American Earth Sciences, (7), 179-207.

Navarro García, L, Hoja Geológico Económica 10 D "Laguna de Aparoma". Provincias de Catamarca y Salta, 88 p. (Unpublished)

Sparks, R., Francis, P., Hamer, R., Pankhurst, R., O'Callaghan, L., Thorpe, R, Page. R., 1985, Ignimbrites of the Cerro Galán Caldera, NW Argentina: Journal of Volcanolgy and Geothermal Research, (24), 205-248.

Turner, J. C., 1962. Estratigrafía de la región al naciente de la Sierra Laguna Blanca, Catamarca: Revista de la Asociación Geológica Argentina, XVII, (1), 11-45.

THE CHACANA CALDERA COMPLEX IN ECUADOR

Minard L. Hall and Patricia A. Mothes
Instituto Geofisico, Escuela Politécnica Nacional
Quito, Ecuador
volcan_pete@yahoo.com

Keywords: caldera formation, tuff, ignimbrite, rhyolite

Abstract

The Chacana caldera complex lies 50 km east of Quito, the capital city, in the Eastern Cordillera of Ecuador's Andes. In order to bring potable water from the eastern slopes of the Eastern Cordillera to the Quito area and the InterAndean Valley, a 20 km-long tunnel is planned that must pass directly through the center of the caldera; this has prompted a detailed geological study of the tunnel alignment. This caldera was suspected and reported upon earlier by Hall and Beate (1991) and Hall and Mothes (1997).

The Pleistocene caldera (3300 - 4200 m elevation) is underlain by a complex metamorphic basement, and both geologic regimes are affected by on-going tectonic activity that includes NE-trending transpressive faulting as well as eastward-trending thrust faulting. The caldera's dimensions are large: the N-S diameter across the entire structure measures 50 km and the E-W width is truncated, but measures at least 30 km (Fig. 1). Outward-sloping packages of ignimbrites, lava flows, and tuffs comprise the north, west, and southwest outer flanks. The caldera rim is irregular in shape and height, but where best developed on the west side it has a height of 500 m. The central depression is everywhere filled, first by ignimbrites, tuffs, and breccias of siliceous nature associated with the volcanism of the outer flanks, and later by post-collapse volcanism consisting of lava emissions of intermediate composition and subsequently by detrital sediments. Post-collapse resurgence began sometime after 0.44 Ma and has uplifted these infilling sediments to heights that define the present continental divide (4200 m). The rhyolitic units associated with early caldera development are 2.5 - 2.7

IOP Conf. Series: Earth and Environmental Science 3 (2008) 012004 doi:10.1088/1755-1307/3/1/012004

Ma in age, while the youngest activity corresponds to two andesitic lavas from the 1700's AD.

Figure 1: The Chacana caldera depression, outlined by heavy bars, is completely filled by post-collapse volcanism and sediment accumulation. The black belt is the fracture zone comprised of many dikes, breccias, and faults. The out-sloping flanks of ignimbrites are identified by arrows. Pambamarca and Las Puntas are pre-caldera andesitic cones, while Antisana is a young, active cone. Intense erosion on the Andes' east flank has left only scarce remnants of the caldera sequence.

The geologic history of the caldera is comprised of four lithologic groups. First and oldest, is the belt of numerous dikes, associated faults, and breccias that form the presumed ring fracture structure along the caldera's eastern margin. The 1 km-wide belt, bounded by metamorphic rocks on the outer east side, is comprised of at least 6 vertical dikes of intermediate and siliceous compositions. On the caldera side of the belt are found two types of breccias exposed as steeply-dipping layers: a breccia made up of andesitic and dacitic clasts in a tuffaceous matrix and a coarse breccia of only andesitic clasts and matrix. Fewer but similar dikes occur at the inner foot of the western caldera rim. The second group corresponds to the caldera's outer flank sequence, more than 1250 m thick, that dips 15-20° away from the caldera rim and depression for as much as 18 km, before disappearing under the InterAndean Valley fill. It is chiefly made up of ignimbrites, welded tuffs, vitrophyres, and tuffs, mostly of dacitic to rhyolitic composition, bearing plagioclase, sanidine, biotite, and quartz as phenocrysts. These are inter-bedded with subordinate lavas of andesitic to dacitic composition. The total sequence ranges in age from 2.7 to 0.8 Ma and has a volume crudely estimated at 670 km^3.

Figure 2: The Potrerillos lava flow (0.26 Ma) with its undulating morphology descended from a resurgent dome vent within the caldera depression. The western caldera rim forms the dark ridge seen in the background, and the outward-sloping western flanks lie behind.

The caldera-infilling makes up the third group, comprised of three rock units: a) below, a thick tuff and breccia unit that has suffered widespread alteration of low to medium intensity, b) a thick series of black porphyritic andesitic lavas dated at 1.5 to 1.9 Ma in age that covered the tuff and breccia unit. These lavas are transitional up section to c) a volcanic sand to gravel to conglomerate succession of fluvial origin that implies progressively more energetic depositional conditions with time and that we interpret to represent the gradual arching of the resurgent area. The fourth group is represented by a long period of lava emissions from a series of vents aligned on a 19 km-long arc that crosses the western half of the caldera depression. Most of these flows have a young morphology, are several kilometers in length, and are siliceous andesites to dacites in composition (Fig. 2). These range in age from 0.45 to 0.16 Ma, however more recent lavas occurred from 30 to 20 ka. The most recent rhyolitic emissions, estimated at 0.22 Ma in age, include major ignimbrites that traveled down valleys on the west flank and out into the InterAndean valley, and a climatic eruption that deposited an 8 m-thick pumice fall unit and a 7 km-long obsidian flow, dated at 0.16 to 0.17 Ma.

In conclusion, although petrologic studies have yet to be completed, the overall Chacana caldera complex has many of the structural and petrographic characteristics of the San Juan Mtns., Colorado, calderas and other rhyolitic calderas. It is part of a grouping of scattered rhyolitic edificies, consisting of both large caldera structures as well as smaller centers, that extend from the Chacana caldera in the north to the Chalupas caldera to the south, a distance of 100 km.

References:

Hall, M. and Beate, B., 1991. El Volcanism Plio-Cuaternario en los Andes del Ecuador. en Mothes, P. (ed) El Paisaje Volcánico de la Sierra Ecuatoriana -Estudios de Geografia, Vol. 4, p. 5-18, Quito.

Hall, M. and Mothes, P., 1997. Chacana Caldera -- The largest rhyolitic eruptive center in the northern Andes: Memorias, 1997 IAVCEI Mtg, México, p. 14.

IOP Conf. Series: Earth and Environmental Science 3 (2008) 012005 doi:10.1088/1755-1307/3/1/012005

Pyroclastic eruptions from Axial Caldera, Juan de Fuca Ridge, NE Pacific Ocean

Christoph Helo [1], Dave A. Clague [2], and John Stix [1]

[1] Department of Earth & Planetary Sciences, McGill University, 3450 University Street, Montreal, Quebec H3A 2A7, Canada

[2] Monterey Bay Aquarium Research Institute

7700 Sandholdt Road, Moss Landing, CA 95039-9644 USA

Keywords: submarine volcanism, caldera, Pele's hair, MORB

Abstract

We present preliminary results of a study on submarine explosive volcanic activity in a mid-ocean-ridge (MOR) environment, Axial Seamount, an active caldera system situated on the Juan de Fuca Ridge in the NE Pacific Ocean. Axial Seamount rises to a depth of 1500 m below sea-level and is characterised by a U-shaped caldera of 3 x 8 km in diameter. Widespread volcaniclastic deposits on its eastern and western flanks testify to the occurrence of explosive activity, in addition to its effusive activity producing sheet and pillow lava flows. Several cores of unconsolidated volcaniclastic deposits were taken from both flanks of Axial Seamount during the MBARI 2006 cruise. The glass fragments comprise a wide diversity of morphologies and textures, including different types of angular fragments, various forms of limu o Pele (bubble wall shards), Pele's hair, ribbon-like forms with tubular vesicles, and sporadic intensively vesiculated clasts (Fig. 1, 2). Angular fragments display surface textures ranging from smooth through cracked to irregular, with or without vesicles. The glass fragments encompass a range of MOR basalt (MORB) compositions. Moderately fractionated magmas with 6.7 – 8 wt% MgO dominate, while more primitive compositions with up to 9.6 wt% MgO are found occasionally. Detailed analysis of a core from the eastern flank suggests that (1) on a vertical dm scale two to three eruption events were at least partially sampled, (2) the chemical homogeneity of the erupted magma varies among the different eruptions, and (3) fragmentation efficiency tends to decreases

IOP Conf. Series: Earth and Environmental Science **3** (2008) 012005 doi:10.1088/1755-1307/3/1/012005

over the duration of one eruptive event, resulting in a coarsening upward sequence of volcaniclastites deposited during one event.

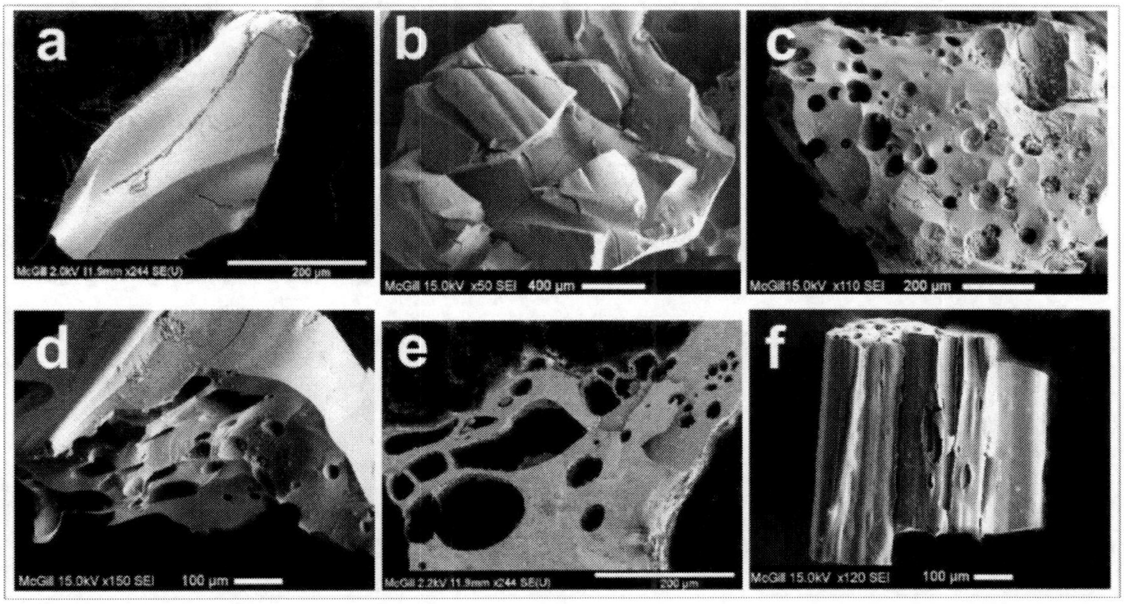

Figure 1: Scanning electron microscope images of various forms of pyroclastic glass fragments from Axial Seamount with different surface textures and vesicularity. **A, b)** blocky fragments without vesicles, **c, d)** vesiculated blocky fragments, and **e, f)** pumiceous fragments with tubular vesicles.

Although fragmentation mechanisms of explosive deep-sea eruptions are still not yet well understood, we speculate that fragmentation of a collapsing foam layer from the top of the magma chamber may account for the production of the various types of pyroclasts. Limu o Pele fragments are thus formed from larger bubbles originating from the collapsing foam, whereas angular pyroclasts may represent the fully collapsed and more degassed and thus vesicle-depleted parts of the collapsing foam. Fragmentation of this mixture is envisioned to be initiated within the conduit, and further disaggregation of the pyroclasts takes place due to rapid quenching of the magma by seawater in the upper level of the conduit and/or upon ejection into the water column. The presence or absence of small vesicles and plagioclase phenocrysts could play a crucial role during this quench fragmentation in determining the final morphology/texture of the angular pyroclasts.

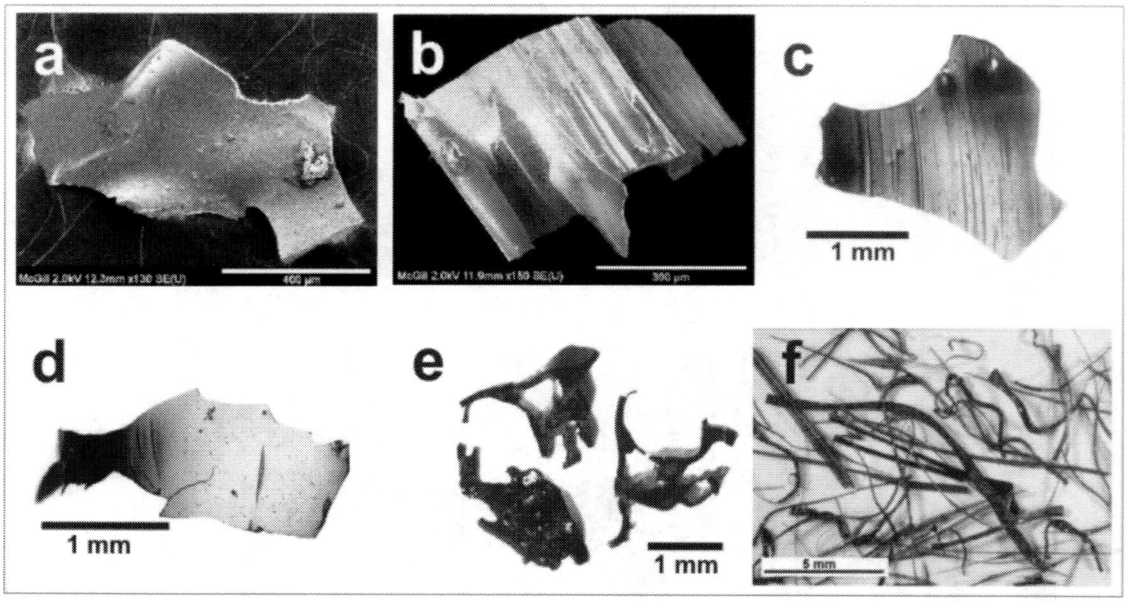

Figure 2: Limu o Pele (bubble wall fragments) and Pele's hair from Axial Seamount. **A, b)** scanning electron microscope images and **c-e)** photomicrograph images of limu o Pele fragments. **F)** photomicrograph images of Pele's hair fragments.

IOP Conf. Series: Earth and Environmental Science 3 (2008) 012006 doi:10.1088/1755-1307/3/1/012006

Physical volcanology of the mafic segment of the subaqueous New Senator caldera, Abitibi greenstone belt, Quebec, Canada

Lyndsay N. Moore* and Wulf U. Mueller

* corresponding author: lyndsay.moore@uqac.ca

Université du Québec à Chicoutimi, 555 boul. du l'Université, Chicoutimi, Québec, Canada, G7H 2B1

Keywords: Archean, caldera, subaqueous, hyaloclastite, ponded magma, lava flow

1. Introduction

Subaqueous calderas are difficult to study because of their inaccessibility. What is known about them has been obtained through research vessels, geophysical instruments and submersible remotely operated vehicles (ROV). Although these methods provide insight into physical surfaces, little is known about internal caldera structures and construction. Alternatively, Archean calderas in the Abitibi greenstone belt, Canada, with vertically dipping strata provide a valuable cross-section of the internal geometry. The Southern Volcanic Zone of the Abitibi greenstone belt hosts the newly proposed Blake River megacaldera complex (Pearson & Daigneault, 2008). The complex is composed of 1) an initial shield building phase known as the Misema caldera (2706-2703 Ma) and 2) two graben-type calderas known as the New Senator (2704-2700 Ma) and the Noranda caldera (2700-2696 Ma; Mueller et al., 2007).

2. Structure

The Misema caldera is a 40 x 80 km E-W trending structure displaying characteristics of a well-developed shield volcano complex comprised of several coalescing volcanoes (e.g. Hawaii and Tenerife). Situated within the inner ring fault of the Misema caldera are the New Senator and Noranda calderas. The New Senator caldera is a 14 x 35 km NW-trending structure overprinted by the NE-trending, 15 x 20 km, Noranda caldera event. Basal sections of the New Senator caldera are mostly mafic in composition and consist of thick massive, brecciated and pillowed units, as well as inferred subaqueous ponded magma units and sills/dykes (Mueller & Corcoran, 2007; Pearson & Daigneault, 2008). Up-section, the sequence is dominated by subaqueous felsic dome-flow complexes (e.g. Glenwood rhyolite). The sequence is cross-cut by numerous felsic and mafic dykes and intruded by the Kiwanis-east tonalite and Kiwanis-west quartz-diorite. New Senator caldera development was accompanied by the generation of caldera-margin and concentric faults which accommodated subsidence events and acted as conduits for hydrothermal fluids (Kerr & Gibson, 1993).

3. Physical volcanology

Our study focuses on the use of physical volcanology to differentiate between ponded magmas, sills and lava flows at the base of the subaqueous New Senator caldera.

Ponded magma: Massive mafic units were interpreted as diorite intrusions or pillowed flows but recent detailed mapping has revealed a series of massive units with hyaloclastite horizons at the inferred tops of the units. Bedding units trend N320 and dip 60° to 80°. Massive units typically range in thickness from 2 to 6 m, with the exception of one large unit approximately 30 m thick near the base of the outcrop. Upper portions of massive units appear aphanitic while toward the base they become subophitic. At the top of each massive division are a series of discrete hyaloclastite units exhibiting a number of primary volcanic textures. Hyaloclastite units range in thickness from 20 cm to 5 m and units become thicker up-section. Contacts between hyaloclastite horizons and massive units are sharp and characterized by a transition from aphanitic magma with few spherulites to abundant spherulites to hyaloclastite (mm to cm scale). Basal contacts of thicker hyaloclastite units exhibit a 1 to 2 m "v-shaped" cone and appear to originate from the underlying massive unit. Small "cigar-shaped" structures are a prominent feature of the hyaloclastite units (Figure 1a). The units exhibit a "pseudo-pillow" form and resemble subaqueous pillowed lava flows. Most hyaloclastite fragments show a preferred orientation indicating rapid lateral movement as the magma was quenched. In thin section, hyaloclastite fragments are angular with polycuspate margins and retain a jig-saw type fit, characteristic of fragments which have not been transported great distances from their source (Figure 1b). Elongated and curvilinear hyaloclastite fragments resemble modern Limu o Pele (Maicher & White, 2001). A series of mafic dykes cross-cut the outcrop at various orientations and a large mafic dyke trending N30° cross-cuts the south-eastern portion. Smaller dykes cross cutting the ponded units show well defined chilled margins and irregular propagation.

We equate these massive ponded units to lavas found in lava lakes of modern subaqueous summit calderas. Hyaloclastite units found along the upper contacts of the ponded magma units are the result of thermal granulation occurring as seawater was ingested into the system. The tops of the ponded magma units interacted with seawater, resulting in vapour phase streaming and the formation of v-shaped segregation pipes, cigar-shaped structures and "pseudo-pillow" forms. Cross-cutting mafic dykes potentially feed the massive ponded magma units up-section and their irregular forms suggest they intruded the ponded units before they were completely consolidated.

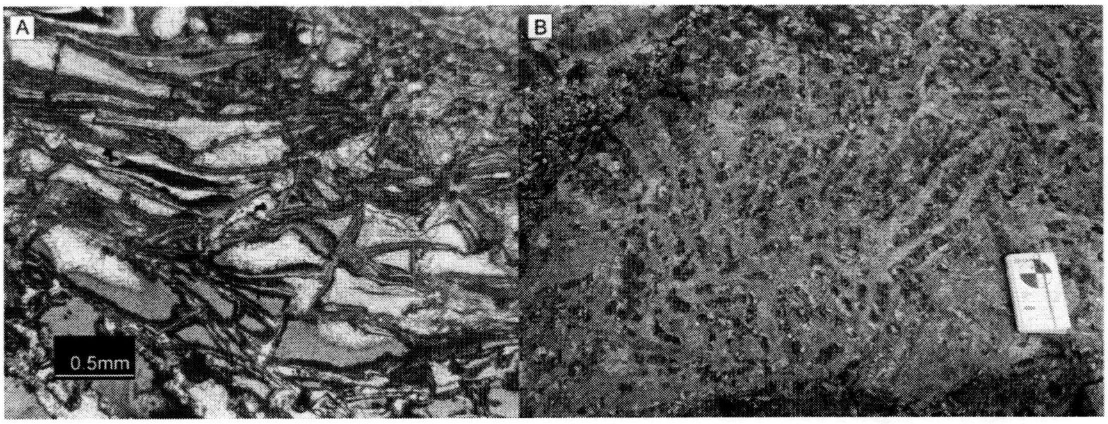

Figure 1. A) Hyaloclastite from ponded magmas of the New Senator caldera. Note close "jig-saw" fit of glass fragments; B) "Cigar" shapes found within hyaloclastite horizons of ponded magmas of the New Senator caldera (notebook is 18 cm).

<u>Mafic lava flows:</u> The central portion of the basal segment of the New Senator caldera is characterized by extensive basalt flows exhibiting classic flow geometry. The E-W striking, 5-20 m thick flows are traceable for several 100's of metres and display both lateral and vertical facies changes. Vertical cross sections of larger flows show a central lava flow with well brecciated margins. Central portions of flows appear massive but upon closer inspection are *in situ* brecciated and contain mm-scale networks of hyaloclastite. Brecciated margins are dominated by hyaloclastite, pillow fragments and pillow rind breccia. Fragments and breccia show imbrication indicating an easterly flow direction and their abundance is characteristic of low-viscosity lava flows. The massive to brecciated flows are bounded to the north and south by pillowed flows, supporting a subaqueous depositional environment and indicating younging to the north. In thin section, hyaloclastite appears significantly altered and epidote is abundant along the margins of the fragments. These flows represent the final mafic portion of the New Senator caldera before felsic flows dominate the volcanic stratigraphy.

Ponded magmas are characterized by dominantly fine grained massive units with distinct hyaloclastite horizons at the top of each unit with v-shaped structures near their base. Hyaloclastite fragments resemble subaqueous Lium o Pele and show a preferred orientation suggestive of vapour phase streaming but do not indicate any mass movement of the ponded units. Mafic dykes and sills intruding into ponded units have well defined chilled margins but irregular forms, suggesting they intruded into the units before they were completely consolidated. Mafic lava flows exhibit classic geometry of low-viscosity flows with central lavas bordered by upper and lower brecciated margins. These brecciated margins are dominated by hyaloclastite, pillow fragments and pillow rind breccia, which are imbricated and indicate flow direction. Differentiation between these elements is important in Archean settings from an exploration point of view because margins of ponded magmas found in modern subaqueous summit calderas are often sites of VMS formation (e.g. Axial Seamount; Hannington & Scott, 1988).

References

Daigneault, R., Mueller, W.U. and Chown, E.H., 2004. Abitibi Greenstone Belt Plate Tectonics: The diachronous history of arc development, accretion and collision, *in* Eriksson, P.G., Altermann, W., Nelson, D.R., Mueller, W.U., and Catuneanu, O. (eds.), The Precambrian Earth: Tempos and Events. Developments in Precambrian Geology, Volume 12, 88-103.

Hannington, M.D. and Scott, S.D., 1988, Mineralogy and Geochemistry of a Hydrothermal Silica-Sulfide-Sulfate Spire in the Caldera of Axial Seamount, Juan de Fuca Ridge. Canadian Mineralogist: 26(3), 603-625.

Kerr, D.J. and Gibson, H.L., 1993, A Comparison of the Horne Volcanogenic Massive Sulfide Deposit and Intracauldron Deposits of the Mine Sequence, Noranda, Quebec: Economic Geology, v. 88(6), p. 1419-1442.

Maicher, D. and White, J.D.L., 2001. The formation of deep-sea Limu o Pele: Bulletin of Volcanology, 63(2), 482-496.

Mueller, W.U., and Corcoran, P.L., 2007, The Abitibi Greenstone Belt: Evolution of an Archean oceanic arc – Field Trip Guide Book. Archean Symposium 2007: A Global Comparison of Archean Terranes. University of Western Ontario, London, Canada: pp. 44.

Mueller, W.U., Pearson, V., and Mortensen, J.K., 2007. Évolution sous-marine du Groupe de Blake River : Nouvelles datation et caractéristiques des caldeiras Archéennes de Misema et de New Senator. Recueil des résumés, Carrefour des Science de la Terre, Univ. Québec à Chicoutimi, p45.

Pearson V. and Daigneault, R., 2008, An Archean megacaldera complex: the Blake River Group, Abitibi greenstone belt: Precambrian Res. In press.

RHYOLITIC CALDERAS AND CENTERS CLUSTERED WITHIN THE ACTIVE ANDESITIC BELT OF ECUADOR'S EASTERN CORDILLERA

Patricia A. Mothes and Minard L. Hall
Instituto Geofisico, Escuela Politecnica Nacional,
Quito, Ecuador
pmothes@igepn.edu.ec

Keywords: caldera, rhyolite, Ecuador, caldera formation

Abstract

The Northern Volcanic Zone of the Andes that extends from central Colombia to central Ecuador is mainly known for its andesitic and dacitic volcanism, related to a well-established subduction system. In Ecuador volcanism manifests itself along three principal rows: the dacitic-andesitic centers of the Western Cordillera, the andesitic stratovolcanoes of the Eastern Cordillera, and the basic alkaline activity of a back-arc group. Many active volcanoes of the Eastern Cordillera have erupted basic andesitic magmas repeatedly during the Late Pleistocene, Holocene, and Historic times.

Relatively unknown, however, is the large volume silicic magmatism that has been occurring since the early Pleistocene in an area 100 km (N-S) by 50 km (W-E) of the Eastern Cordillera, that lies 30-60 km east and southeast of Quito, Ecuador's capital (Fig. 1). It is comprised of a cluster of scattered rhyolitic and dacitic centers that includes both large caldera structures as well as smaller edificies, that are built upon a Paleozoic- Mesozoic metamorphic basement. This magmatism dates from 2.7 Ma at Chacana to almost historic times at Cotopaxi.

At least seven distinct rhyolitic centers are recognized. The large Chacana caldera complex, containing many vents of andesitic to rhyolitic composition, is the most northern vent of the silicic cluster (see Hall and Mothes, this volume). At the southern end lies the 15 x 20 km-diameter Chalupas Caldera with its typical caldera morphology, displaying a well-developed caldera rim, a flat interior depression, resurgent features, and out-flowing deposits.

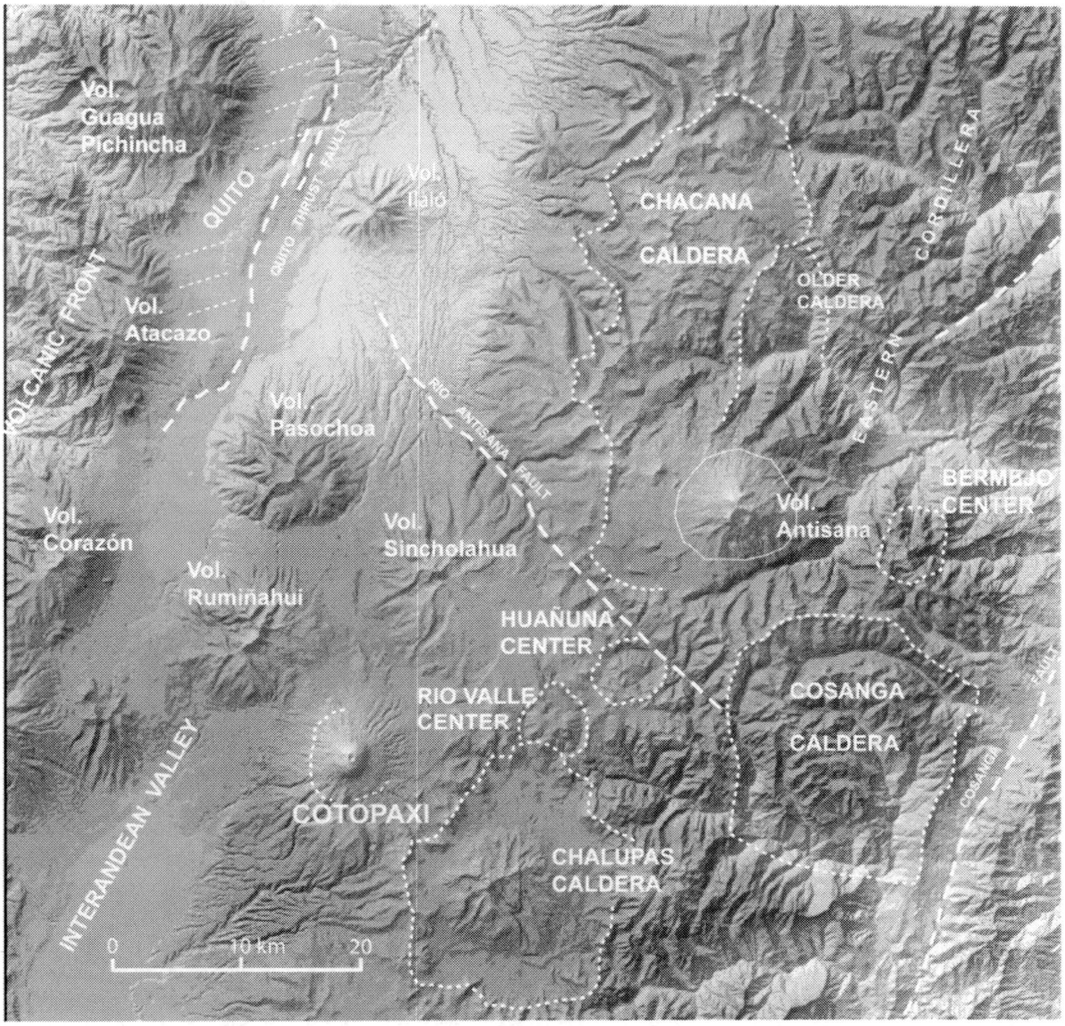

Figure 1. Regional map of North-Central Ecuador showing all known rhyolitic centers of the Eastern Cordillera. Quito, Ecuador's capital, lies 30 to 60 km to the west and northwest of these centers.

Andesitic and dacitic lava flows and vents form the pre-caldera edifice. The caldera's climatic eruption 211 ka ago produced >100 km^3 of rhyolitic pumice flows (Hammersley et al., in press) which resulted in the foundering and collapse of the caldera (Beate, 1989). These flow deposits fill much of the InterAndean Valley depression (>3000 km^2) and are also found at the eastern and western foot of the Andes. Resurgent activity was dominated by the Quilindaña stratocone and associated vents. Nearby Cotopaxi volcano has built its andesitic cone upon the remnants of a ~0.5 Ma old rhyolitic caldera, however following an extended repose interval it again erupted rhyolitic ash-flows and tephra on seven occasions between 13 and 2.2 ka (Hall and Mothes, 2007).

Recent fieldwork has identified two young eruptive centers located to the north of the Chalupas caldera. The Rio Valle Center is comprised of several rhyolitic domes and draping deposits of pyroclastic flows that also fill the nearby Yana-urcu and Valle river valleys. The pumice-rich flow deposits contain light grey obsidian, grey-banded rhyolite lava clasts, and white, biotite-bearing, tubular-vesiculated pumice (Fig. 2). Based upon tephra and glacial stratigraphy its age is placed between 15 and 6.3 ka.

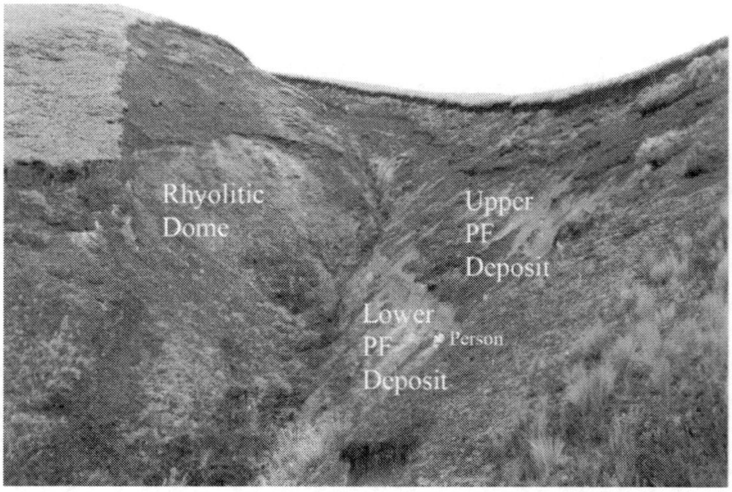

Figure 2: Rio Valle dome and associated ignimbrites that drape the dome and traveled several kilometers down valley. Note person for scale.

Five km further to the northeast lies the Huañuna center, also thought to be late Pleistocene to early Holocene in age. It appears to have a resurgent dome center, encircled by glacial valleys that contains its youngest ash-flow deposits, and farther afield, by outward-sloping ignimbrites (Fig. 3). Late glacial moraines are covered by its pyroclastic flow deposits. The chemistry of its pumice is nearly identical to that of Chalupas' main pyroclastic flow. Preliminary studies suggest the intrusion of Chalupas-like rhyolitic magmas into nearby centers within 15 km of its caldera. Garrison et al. (2003) argued for the recycling of Chalupas magma that initiated the 4.5 ka rhyolitic eruption of Cotopaxi volcano.

To the northeast of Chalupas, another large circular structure exists, named Cosanga, but its inaccessibility and dense cloud forest have prevented further study.

IOP Conf. Series: Earth and Environmental Science 3 (2008) 012007 doi:10.1088/1755-1307/3/1/012007

Figure 3: The Huañuna Dome lies at the center of a circular feature, and is flanked by outward-sloping ignimbrites.

Nevertheless, Cosanga's shape and proximity to other rhyolitic centers would suggest that it is genetically-related. Immediately to its north, the Bermejo center is known as a source of rhyolitic obsidian flows and ash-flow deposits found in the Cosanga Fault valley.

This cluster of silicic centers overlies the most negative gravity anomalies identified along the crest of the NVZ (Feininger and Seguin, 1983) and crustal thickness under this region is estimated at 50-70 km (Gullier et al., 2001). Such thicknesses would imply higher temperatures and suggest crustal assimilation. However, based upon strontium isotopic data from Cotopaxi and Chalupas, both Garrison et al. (2003) and Hammersley et al. (in press) concluded that only very limited crustal assimilation had occurred and that most rhyolites were generated primarily by differentiation processes.

References:

Beate, B., 1989. The Chalupas ignimbrite. Abst IAVCEI Gen Assembly-New Mexico. New Mexico Bureau of Mines & Mineral Resources Bull 131.

Feininger, T., Seguin, C.R., 1983. Simple Bouguer gravity anomaly field and the inferred crustal structure of continental Ecuador, Geology, 11, pp. 40-44.

Garrison, J., Davidson, J., Turner, S., Reid, M., 2003. Recycling of the Chalupas pluton at Cotopaxi volcano, NVZ, Ecuador: evidence from 238U-230Th disequilibria. Geophy Res Lett Abst 5:11998

Guillier, B., Chatelain, J.L., Jaillard, E., Yepes, H., Poupinet, G., Fels, J.F., 2001.

Seismological evidence on the geometry of the orogenic system in central-northern Ecuador (South America), Geophy Res Lett, Vol. 28, No. 19, pp. 3749-3752.

Hall, M., Mothes, P., this volume. The Chacana Caldera Complex -- Ecuador.

Hall, M., Mothes, P., 2007. The Rhyolitic-Andesitic Eruptive History of Cotopaxi Volcano, Ecuador. Bull. Volcan. DOI 10.1007/s00445-007-0161-2.

Hammersley, E., DePaolo, D., Beate, B., and Deino, A., in press, Quantitative model for andesite and rhyolite generation in a continental margin volcanic arc: the Chalupas caldera, Ecuador;.

IOP Conf. Series: Earth and Environmental Science 3 (2008) 012008 doi:10.1088/1755-1307/3/1/012008

Geology and structure of the Malpaso caldera and El Ocote ignimbrite, Aguascalientes, Mexico

Nieto-Obregón, Jorge (Facultad de Ingeniería, UNAM, Coyoacán, 04510, México D.F. MEXICO; nieto@servidor.unam.mx)

Aguirre-Díaz, Gerardo (Centro de Geociencias, UNAM, Campus Juriquilla; 76220, Querétaro, Qro., México; ger@geociencias.unam.mx)

Keywords: caldera, ignimbrite, faults, graben

1. Introduction

A new caldera is reported west of the city of Aguascalientes, Mexico (Figure 1). The caldera is a volcano-tectonic depression, highly fractured and faulted, and was filled by voluminous pyroclastic products related to the caldera collapse. Due to these characteristics, Aguirre-Díaz et al. (2008) classify it as a graben caldera. It is truncated by younger normal faults of the Calvillo and Aguascalientes grabens (Nieto-Obregón and Aguirre-Díaz, 2007a). In this work we present a summary of the geologic and structural observations on this caldera.

2. Stratigraphy.

The general stratigraphy in the caldera area consists of: 1) a Mesozoic metamorphic basement, 2) a continental sedimentary sequence (red beds) that is intercalated with pyroclastic surge deposits and non-welded ignimbrites erupted in the caldera's earlier stage, 3) The El Ocote Ignimbrite, the main product of the caldera, 4) rhyolitic domes that represent post-collapse effusive eruptions of highly viscous rhyolitic magma, which in turn are related to other ignimbrites and pyroclastic breccias produced by flank collapse of the domes and 5), andesitic feeder dykes and associated flows that intrude and cover all the caldera sequence (Figure 2). Normal faulting associated with Calvillo and Aguascalientes grabens facilitated the emplacement of these dykes and lavas.

IOP Conf. Series: Earth and Environmental Science 3 (2008) 012008 doi:10.1088/1755-1307/3/1/012008

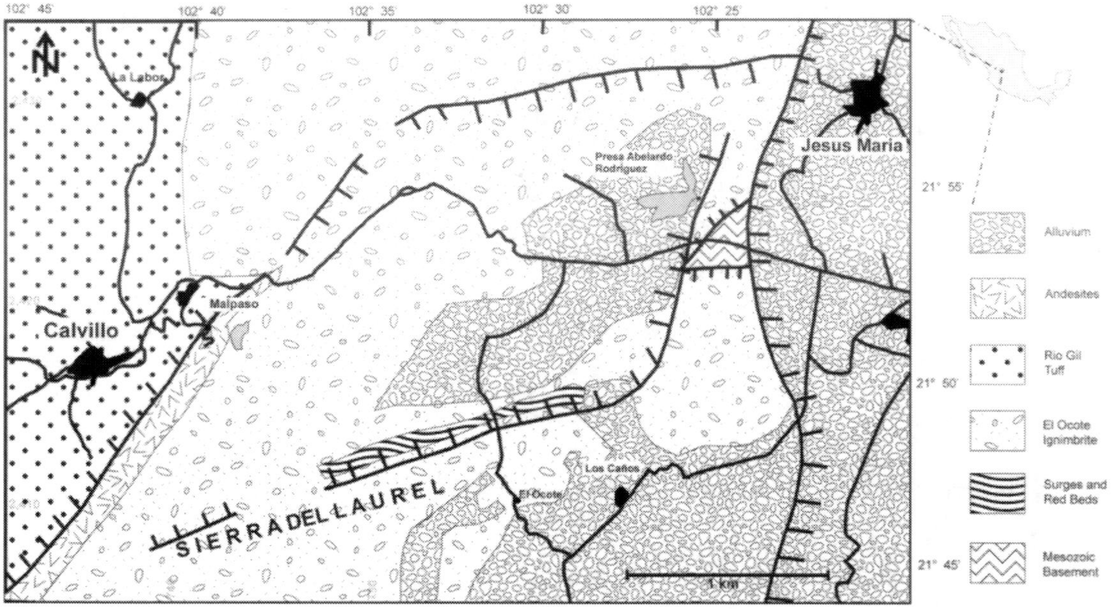

Figure 1. Simplified geologic map of the Caldera of Malpaso showing the major structures and main lithologic units.

The El Ocote Ignimbrite fills the caldera depression and extends several km outwards from the caldera. The intra-caldera facies consist of a high grade reomorphic ignimbrite with lava-like features, including flow banding and folds. The pyroclastic nature of this unit is demonstrated by the presence of a eutaxitic fabric with abundant elongated fiammes and cm-size lithics rotated within the flow foliation.

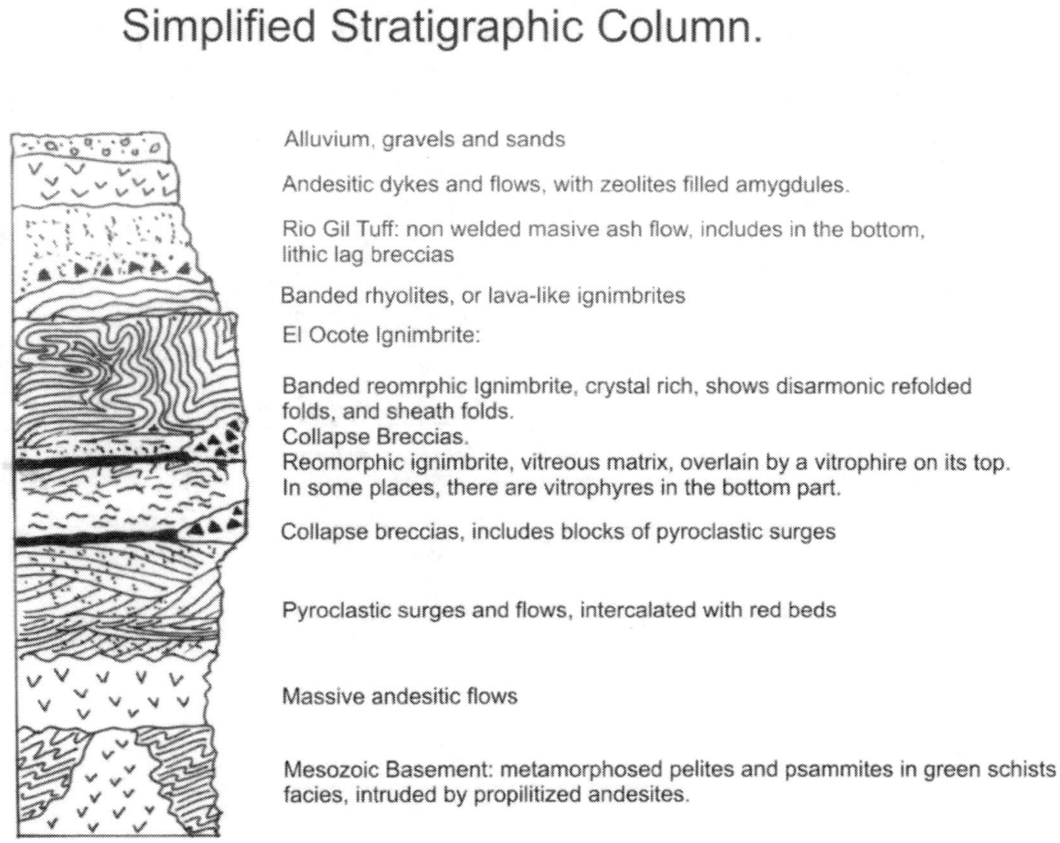

Figure 2. Simplified stratigraphic column and description of main units.

Devitrified glass shards and axiolites of collapsed pumices are observed under the microscope. The glass shard matrix is deformed by flow around rigid phenocrysts (Figure 3). This ignimbrite is considered as reomorphic because there is evidence of ductile deformation of hot pyroclastic material forming folds of different styles that include gentle, open, close, tight, isoclinal and sheath type. The open folds are often formed from refolded folds of the foliation. Structural elements such as foliation, axial planes, fold axis and elongation lineations, suggest deposition over an irregular surface, flowing in different directions, from various sources within and in the borders of the caldera (Nieto Obregón and Aguirre Díaz, 2007b). The reomorphic ignimbrite is mainly distributed in the central portion of the caldera and although its thickness has not been observed complete, more than 300 m are exposed in the southern border of the caldera in the Sierra El Laurel. The extra caldera facies of this unit has thinner thicknesses, from 20 to 150 m, observed in the Mesa san José de Gracia, and to the south of Sierra El Laurel.

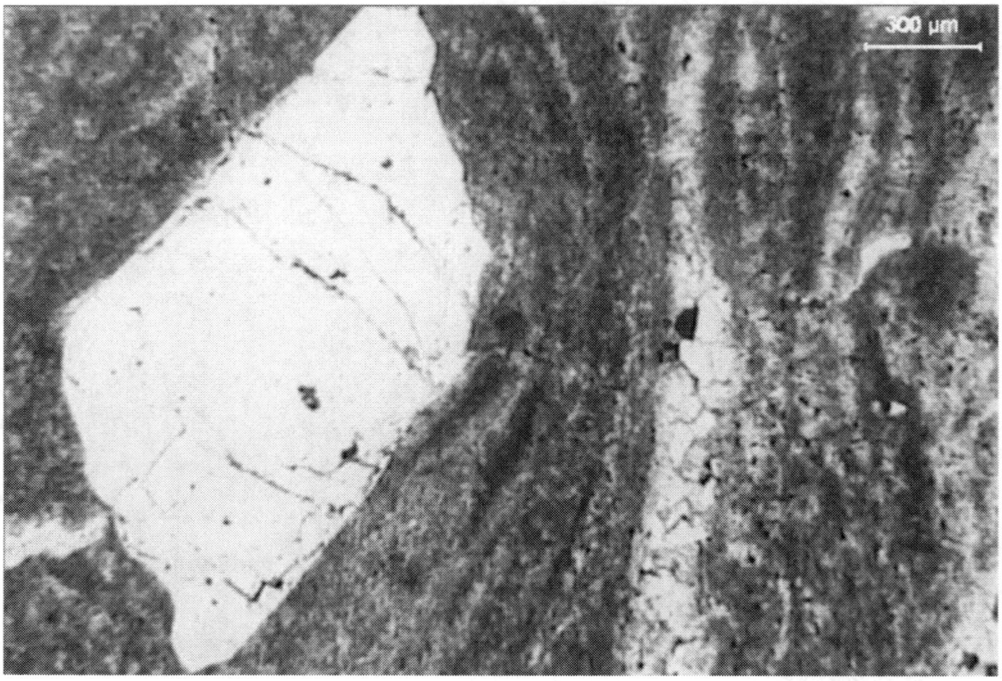

Figure 3. Photomicrograph of El Ocote Ignimbrite. Note the deformation of the glass shards due to intense welding around a feldspar phenocryst.

3. Structure

The area is affected by two types of faults of different ages: 1) volcano-tectonic faults, produced by caldera subsidence processes and 2), tectonic faults of post-caldera age that truncate the western and eastern borders of the caldera. To the east, the caldera is cut by the western fault system of the Aguascalientes Graben, and to the west by the eastern fault system of the Calvillo Graben. From the caldera's fault data and stratigraphy we conclude that the caldera formed by the chaotic subsidence collapse of the magma chamber roof in a piece-meal style, and thus we classify it as a piece-meal graben caldera (*sensu*: Aguirre-Díaz et al., 2007).

References.

Aguirre Díaz, GJ, Labarthe-Hernández, G, Tristán-González, M, Nieto-Obregón, J and Gutiérrez-Palomares, I., 2007, Graben Calderas. Volcano-tectonic explosive collapse structures of the Sierra Madre Occidental, Mexico. Geophysical Research Abstracts,

Vol. 9, 04704, S-Ref-ID: 1607-7962/gra/EGU2007-A-04704. European Geosciences Union, 2007.

Aguirre-Díaz, G.J., Labarthe-Hernández, G., Tristán-González, M., Nieto-Obregón, J., Gutiérrez-Palomares, I., 2008, Ignimbrite Flare-up and graben-calderas of the Sierra Madre Occidental, Mexico: in J. Martí and J. Gottsmann, eds., "Caldera Volcanism: Analysis, Modelling and Response", Developments in Volcanology 10, Elsevier, Amsterdam, 492 p., (143-180 pp.).

Nieto-Obregón, Jorge and Aguirre-Diaz, Gerardo, 2007a, Geología y Estructura de la caldera de Malpaso, Aguascalientes, Mexico. Geos, Vol. 27, no.1, p. 131-132.

Nieto-Obregón, Jorge and Aguirre-Diaz, Gerardo, 2007b, Ignimbrita El Ocote. Una ignimbrita de muy alto grado en la caldera de Malpaso, Aguascalientes, México. Mexico. Geos, Vol. 27, no.1, p. 130.

NEW TOOLS TO INVESTIGATE TEXTURES OF PYROCLASTIC DEPOSITS

[1*]Damiano Sarocchi, [2]Lorenzo Borselli and [3]José Luis Macías

[1]Instituto de Geología / Fac. Ingeniería UASLP, Dr. M. Nava No 5, Zona Universitaria 78240, San Luis Potosí, Mexico. *e-mail: damiano.sarocchi@uaslp.mx
[2]Istituto di Ricerca per la Protezione Idrogeológica, Consiglio Nazionale delle Ricerche, Via Madonna del Piano 10, 50019, Sesto Fiorentino, Italia.
[3]Departamento de Vulcanología, Instituto de Geofísica, Universidad Nacional Autónoma de México, Coyoacán 04510, D.F., México.

Keywords: textural analysis, image analysis, pyroclastic flows

1. Introduction

Pyroclastic flows are a difficult subject to study due to its dangerousness, difficulties to approach them and the rapidity with which they develop and settle. The few films taken of these flows from short distances revealed poor information because they are involved by an impenetrable ash cloud. In spite of the rapid development of technology (satellites, high speed and thermal cameras, infrasonic sensors, etc.) most information of these flows is still obtained by studying the deposits, scaled experiments and numerical models.

A pyroclastic flow deposit (primary or reworked) keeps a fingerprint of the physical processes that occurred in the flow during transport and settling. This information is recorded in the texture and stratigraphy of the deposit offering an instantaneous view of the flow prior to freezing. Pyroclastic flow deposits reflect characteristics inherited from the source, fragmentation style, segregation and bulking phenomena, which occurs during transport and deposition.

With the term textural properties we refer to the characteristics (bulk or particulate) related with the external geometry and the reciprocal spatial arrangement of the particles constituting a sedimentary deposit. The main textural properties of such a deposit are: granulometry, clasts shape and fabric.

From long time, textural analysis has been used by geologists to infer the sedimentological characteristics of the deposits and the physical properties of the parental flows. However, the attempts to quantify such properties are a difficult task due to the tedious and time consuming procedure that limited its applicability. For this reason, in geology, it was a common practice to quantify the textural properties of sediments by the use of comparative charts (Powers, 1953; Harrell, 1984), even though, the information obtained was semi-quantitative and hinged by human bias. Nowadays, the rapid growing of computer power and the advent of image processing techniques allowed the used of old textural methods with the acquisition of large data sets in short periods of time and statistically sounded. In this work, we introduce some texture's analysis techniques that take advantage of computer processing time.

2. Granulometry

Some important problems are always present when performing granulometric studies in pyroclastic deposits: 1) often, the clast size range is so wide (spans from microns to many meters) that is impossible to cover all the classes with the same analytical method; 2) In many cases, the outcrop is inaccessible for sampling or 3) the deposit is indurated and the particles cannot be separate for sieving. All these problems can be solved using optical granulometry a very versatile tool based on image analysis and stereological techniques, as it is explained below:

Rosiwal's intercepts method provides stereologically correct volumetric data from a scaled photograph (Chayes, 1956; De Hoff and Rhines, 1968; Sarocchi et al., 2005). The method consists in superimpose a set of lines (intercepts) to the outcrop image and to measure the length of segments stacked over each clast. The ratio between stacked segments and the total line length provides the volumetric percentage of clasts. The method is useful to study the coarse components of pyroclastic or epiclastic deposits, completing the granulometric information obtained with the sieving and laser techniques in order to acquire the complete granulometric analysis of the outcrop (***Total Granulometric Analysis***, TGA). This technique is also useful to obtain the granulometric spectrum of cemented or welded deposits. The Rosiwal's intercepts method can also be used to carry out ***Vertical Granulometric Profiles*** (VGP) of medium to coarse size clasts (Sarocchi et al., 2005).

A common problem in granulometric analysis is the acquisition of ***statistical parameters of polymodal distributions***. Inman (1952) and Folk & Ward (1957) methods notoriously cannot be correctly used in samples that show a string non-Gaussian size distribution (e.g. polymodal and/or skewed). For this reason, we wrote a computer program (DECOLOG 2.0, Borselli and Sarocchi 2006, www.decolog.org), that decodes the information present in samples with polymodal distributions, as paradigm, the 3-parameters log-normal distribution and, in particular searching an optimal mixture of these distributions. Besides providing statistical parameters of each log-normal component, the DECOLOG software provide Folk and Ward equivalent parameters optimized by means of a Monte Carlo interpolation (Sarocchi, 2006).

Shape analysis: Particle's morphology is an important textural property that records the sedimentological history of the particle at different scales. Shape is a hierarchical property (Barrett, 1980) with different parameters (form, roundness and surface texture). To quantify such property, we use different methods: Fourier shape analysis (Schwarcz and Shane, 1969) and fractal shape analysis (Orford and Whalley, 1983). Both methods are applied to particles perimeters extracting 2D shape information. ***Fourier shape analysis*** develops the perimeter of the particle with respect to the polar coordinates (Figure 1) obtaining a Geometric Signature Waveform (GSW). The GSW can be analyzed by means of the Fast Fourier Transform as any periodic wave, obtaining phase and amplitude information of the constituent harmonics. Each harmonic provides information related to certain irregularity scale (lower harmonics are related with coarse irregularities and higher harmonics are related with finer irregularities). The amplitude of each harmonic corresponds to the "weight" of the related scale of irregularities. We proposed some morphological coefficients (MC) based on the harmonic amplitudes that are sensible to different shape scales, quantifying the importance of the different shape orders in each particle.

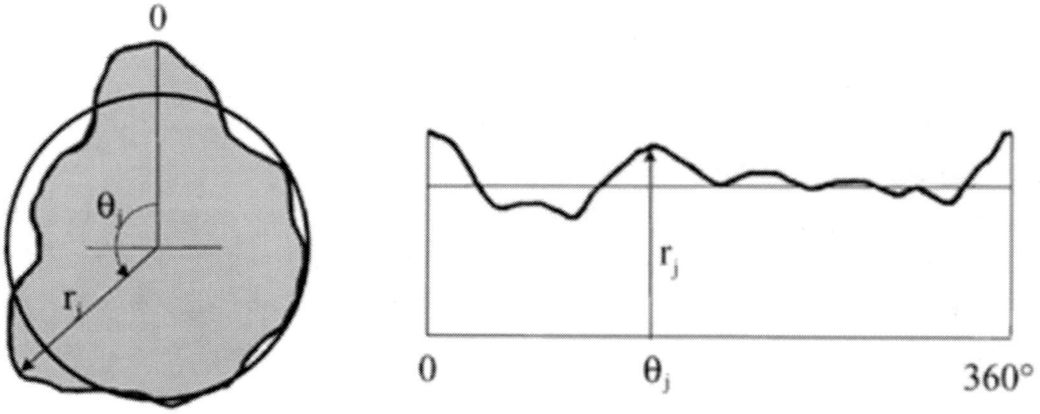

Figure 1. Example of Geometric Signature Waveform (GSW) obtained unrolling the particle's perimeter with respect to the polar coordinates obtaining a Geometric Signature Waveform (GSW). The GSW can be analyzed by means of the Fast Fourier Transform as any periodic wave obtaining information about the constituents' harmonics.

Fractal geometry, synthesized by Mandelbrot (1977), against Euclidean geometry, admits fractionary topological dimensions. This means that certain lines (topologic dimension 1)

constitute so irregular curves that fill the bidimensional space trending to the topological level 2, typical of surfaces. The degree of irregularity of a line is quantified by means of fractal geometry and different techniques to evaluate if a certain tendency exists. We used the "caliper method" (Orford and Whalley, 1983) which approximates the irregular perimeter of a particle with polygons with a progressively larger number of sides (n). Knowing the segment length and the correspondent measured perimeter, we calculate the fractal dimension (D). As the D value increases the perimeter irregularities also increase.
The Fractal geometry and Fourier analysis are useful and precise tools that only describe the bi-dimensional form. In order to describe the three-dimensional general form of a particle, it is necessary to measure its three main axes. Measuring particle's axes, is a time consuming processes, because a large number of measurements is needed to obtain reliable data. Moreover very small particles are impossible to manage and measure. For this reason we improved a method (***Shape From Shading***, SFS), originally proposed for Kaye (1999), which obtains the third dimension of particles by taking a picture of the particle and the projected shadow. By using image analysis, we are able to obtain accurate and quick measurements of the three axes of many particles. The method can be used to measure very small particles by taking photographs with a microscope.
Fabric: With the term fabric we define the three-dimensional arrangement of the particles constituting a sedimentary deposit. Elongated particles suffer the field force originated during the interaction with other particles within the flow. Such particles aligned themselves in order to minimize these forces. Although different conditions exists within a flow (Jeffery, 1922; Lindsay, 1968) a common process is that particles arrange themselves with the long axes (a) parallel to the flow direction (Sestini and Pranzini, 1964; Best, 1992; Capaccioni and Sarocchi, 1996). In order to obtain fabric information we collect an oriented sample of the outcrop (Prior et al., 1987) that is cut in horizontal slices through the middle part of the sample. In these surfaces, the axes (a) of the elongated particles are measured with respect to the north by means of image analysis. To obtained preferential orientations of the particles we used circular statistics and confidence intervals of the whole particle population (Capaccioni et al., 1997). Once, preferred orientations are found, we sectioned the other half of the sample along the preferred azimuthal plane (perpendicular to the horizontal planes) in order to obtain information of particle imbrication. The method is useful to analyze natural or artificially cemented sediments. In natural samples it allows to recognize paleoflows directions and provides insights on the rheological characteristics of the deposits.

3. Conclusions

All the textural analysis described constitute a useful toolkit to study the textural properties of pyroclastic deposits of any volume and type (from ignimbrite to block and ash flow deposits), epiclastic deposits (debris flows, hyperconcentrated flows and normal stream flows) as well as other kind of sedimentary deposits, rocks or lavas.

References

Barrett, P.J., 1980, The shape of rock particles, a critical review: Sedimentology, 27, 291-303.

Best, J.L., 1992, Sedimentology and event timing of a catastrophic volcaniclastic mass flow, Volcan Hudson, Southern Chile: Bull. Volcanol., 54: 299-318.

Borselli. L. and Sarocchi D. 2006. ***Deconvolution of mixtures of lognormal components From particle size distributions*** . http://www.decolog.org/decolog_manual.pdf

Capaccioni, B. and Sarocchi, D., 1996, Computer-assisted image analysis on clast shape fabric from the Orvieto-Bagnoregio ignimbrite (Vulsini District, central Italy): implications on the emplacement mechanisms: Journal of Volcanology and Geothermal Research, 70, 75-90.

Chayes, F., 1956, Petrographic modal analysis: Wiley, New York. 113 p.

DeHoff, R.T., Rhines, F.N., 1968, Quantitative Microscopy: McGrow-Hill, New York. 422 p.

Folk, R.L., Ward , W.C., 1957, Brazos River bar: A study in the significance of grain size parameters: Journal of Sedimentary Petrology, 27(1), 3-26.

Harrell, J., 1984, A visual comparator for degree of sorting in thin and plane sections: Journal of sedimentary research, 54 (2), 646-650.

Kaye, B.H., 1999. Characterization of Powder and Aerosols: Wiley-VCH, New York, 312 p.

Inman, D.L., 1952, Measures for describing the size distribution of sediments: Journal of Sedimentary Petrology, 22(3), 125-145.

Lindasy J.F., 1968, The development of clast fabric in mudflows: Jour. Sed. Petrology: 38: 1242-1253.

Mandelbrot B.B., 1977. Fractals: form, chance and dimension: W.H. Freeman, San Francisco. 361 p.

Jeffery, G.B., 1922, The motion of ellipsoidal particles immersed in a viscous fluid: Royal Soc. London Proc.:102 (A): 161-179.

Orford, J.D. and Whalley, B., 1983. The use of fractal dimension to quantify the morphology of irregular-shaped particles: Sedimentology, 30: 655-668.

Powers, M.C., 1953, A new roundness scale for sedimentary particles: Journal of Sedimentary Petrology, 23 (2): 117-119.

Sarocchi, D., 2006, Análisis textural del depósito de flujo de bloques y ceniza del 17 de Julio de 1999 en el Volcán de Colima: Ciudad de México, Universidad Nacional Autónoma de México, tesis doctoral, 222 p.

Sarocchi, D., Borselli, L., Macías, J.L., 2005, Construcción de perfiles granulométricos de depósitos piroclásticos por métodos ópticos: Revista Mexicana de Ciencias Geológicas, 22 (3), 371-382.

Schwarcz, H.P., Shane, K.C., 1969, Measurement of particle shape by Fourier analysis: Sedimentology, 13: 179-212.

Sestini, G., Pranzini, G., 1964, Orientamento dei granuli e delle impronte di fondo in due formación torbiditiche della Toscana: Boll. Soc. Geol. Ital.:83: 275-290.

Prior, D.J., Knipe, R.J., Bates, M.P., Grant, N.T., Law, R.D., Lloyd, G.E., Welbon, A., Agar, SM., Brodie, K.H., Maddock, R.H., Rutter. E.H., White, S.H., Bell, T.H., Ferguson, C.C. and Wheeler, J., 1987, Orientation of specimens: Essential data for all fields of geology: Geology, 15: 829-831.

Unravelling the collapse mechanisms at a Jurassic caldera of the Chon Aike silicic LIP in Southern Patagonia (47° 15′S, 71° 40′W), Argentina

Sruoga, P.[1], Japas, S.[2], Salani, F.[2] Kleiman. L.[3] and Graffigna, M.[4]

[1] CONICET-SEGEMAR. Av. J. A. Roca 651. Buenos Aires, Argentina.
patysruoga@yahoo.com.ar
[2] CONICET-UBA. Depto. Cs. Geológicas, Pabellón II, Ciudad Universitaria, C1428EHA, Buenos Aires, Argentina. msjapas@gl.fcen.uba.ar, fms@gl. fcen.uba.ar
[3] Gerencia de Exploración de Materias Primas, Comisión Nacional de Energía Atómica, Avda. del Libertador 8250, 1419, Buenos Aires, Argentina. kleiman@cae.cnea.gov.ar
[4] Universidad Nacional de San Juan

Keywords: collapse, megabreccia, ignimbrites, Jurassic, Southern Patagonia

Abstract:

La Peligrosa Caldera is located at Sierra Colorada (47° 15′S, 71° 40′ W) in the Chon-Aike silicic Large Igneous Province (Pankhurst *et al.*, 1998). This volcanic province has an area of 1.7 x 10^6 km^2, including the continental platform and the Antarctic Peninsula with an estimated volume of 235,000 km^3 Middle to Late Jurassic in age (188-153 Ma, Pankhurst *et al.*, 2000), this long-lived and widespread volcanism took place in Patagonia along the western margin of Gondwanaland. Remarkably homogeneous in composition (Sruoga, 1989; Pankhurst *et al.*, 1998), it is dominated by voluminous ignimbritic plateaus, with granites, lava domes, minor intermediate lavas and epiclastic tuffs. However, collapse calderas have been exceptionally recognized. La Peligrosa Caldera (Sruoga, 1994, 2002) represents an unique window to understand the eruptive mechanisms that prevailed throughout the ignimbritic flare-up in Southern Patagonia during middle to late Jurassic times.

On the northern flank of Sierra Colorada, Tertiary Andean thrusting and intense glacial erosion have helped to expose the roots of the deeply dissected La Peligrosa Caldera. Key pieces of lithologic and structural evidences are taken into account to reconstruct the volcanic structure (Sruoga et al., 2008).

The sequence outcropping between Cerro Ghío and La Peligrosa ranch (Figure 1) allowed us to unravel the collapse mechanisms leading to the caldera formation. Three units may be differentiated in this sequence: La Salina Breccia, Los Acantilados Ignimbrite and Cerro Ghío Ignimbrite (Figure 2).

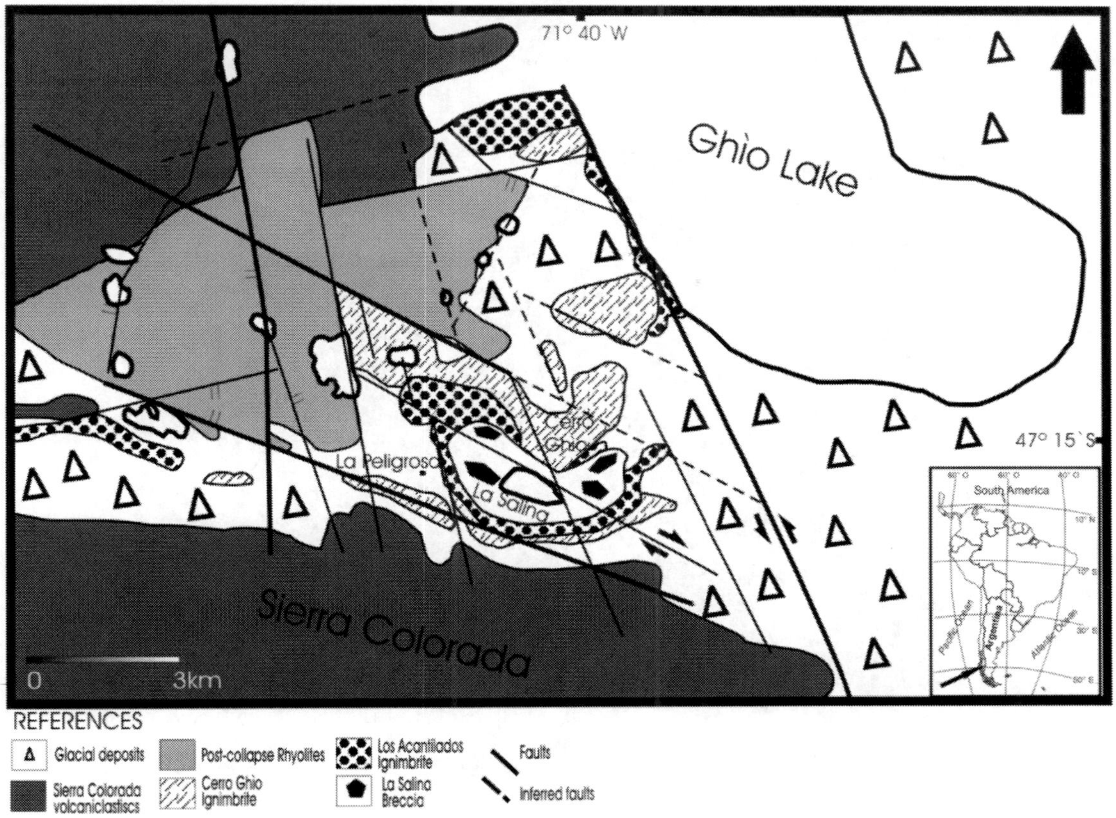

Figure 1: Geologic map

At the base, La Salina Breccia constitutes a chaotic deposit, *ca.*100 m thick, with blocks up to 300 m across, immersed in a pyroclastic matrix. The breccia includes different types of lithics: Paleozoic basement schists, phyllites, quartzites and granites, Jurassic (?) andesites, rhyolites and ignimbrites, and laminated and deformed lacustrine sediments. This mega- to mesobreccia (Lipman, 1976) thins out to the west, accompanied by strong variation in lithic size and content. In the upper part, a transition to the overlying unit may be recognized, based on lithic depletion and strong size reduction. The Los Acantilados Ignimbrite may be described as poorly to moderately welded lithic ignimbrite. This *ca.* 300 thick whitish and massive deposit is characterized by a large proportion of lithics, up to 0.50 m in size. It shows vertical zonation, with green fiamme layers developing eutaxitic textures. In the upper part of this deposit two lithic-rich layers are recognized.The uppermost unit is the Cerro Ghío Ignimbrite. It corresponds to a *ca.* 400 meters thick, massive, dark green, crystal-rich and densely welded cooling unit.It shows an order of magnitude decreasing thickness at the suspected caldera margins. South and westwards, the ignimbritic deposit thins to *ca.* 40-60 m.

Figure 2- Northwestern flank of Sierra Colorada. LSB: La Salina Breccia (C: lithic clasts), LAI: Los Acantilados Ignimbrite, CGI: Cerro Ghío Ignimbrite

The described section records at least two caldera collapse events. The first one is represented by the La Salina Breccia. The megabreccia at the bottom and the vertical and lateral variation in lithic size and concentration suggest that the collapse cannot be described as a single piston-like event. The lithic concentration levels at the upper part of Los Acantilados represent minor local incremental adjustments during the rapid collapse. The emplacement of this large-volume lithic ignimbrite records the initial filling of the caldera. Shortly afterwards, due to the absence of reworked epiclastics, the collapse continued with the reactivation of normal faults and the emplacement of the Cerro Ghío Ignimbrite. Rhyolitic lava domes and flows correspond to post-collapse volcanism.

A prevailing transtensional regime controlled the eruption of the silicic large-volume ignimbrites and lava flows. La Peligrosa Caldera formed after a dilatational zone related to a releasing step-over associated with a left hand *en-échèlon* array of N-S to NNW regional faults. Caldera collapse is dominated by NNW sinistral and WNW dextral transtensional faults (Fig.1). Overprinting caldera structures, and later ENE extensional faults are linked to the rhyolitic post-collapse volcanism and indicate a temporal change in stress conditions.

The relationship between volcanic facies (megabreccia and two intracaldera ignimbrites) and main normal faults allow us to propose a piecemeal-trapdoor type caldera, with the deepest part located between Cerro Ghío and the northwestern flank of Sierra Colorada. Although many pieces are lacking to fully support this model, future studies will provide new elements for better understanding the collapse mechanisms of La Peligrosa Caldera and test its application to the rest of the Chon Aike silicic LIP.

REFERENCES

Lipman, P.W., 1976. Caldera-collapse breccias in the western San Juan Mountains, Colorado: Geological Society of American Bulletin, v.87, p. 1397-1410.

Pankhurst, R. J., Leat, P. T., Sruoga, P., Rapela, C. W., Márquez, M., Storey, C. and Riley, T.R. 1998. The Chon-Aike province of Patagonia and related rocks in West Antarctica: a silicic large igneous province. J. Volcanol. Geotherm. Res. 81, 113-136.

Pankhurst, R. J., T. R. Riley, C. M. Fanning and S. P. Kelley, 2000. Episodic Silicic Volcanism in Patagonia and the Antartic Peninsula: Chronology of Magmatism associated with the break-up of Gondwana. J. Petrol. 41, 605-625.

Sruoga, P.,1989. Estudio petrológico del plateau ignimbrítico jurásico a los 47° 30´L S. Unpublished Ph.D. thesis. Universidad Nacional de La Plata, 400 p.

Sruoga , P., 1994. El Complejo Caldera La Peligrosa. Cordillera Patagónica Austral (47` 15' S). 7° Congreso Geológico Chileno, Actas 2:1219-1223, Concepción

Sruoga, P., 2002. La Peligrosa Caldera: a key for the Jurassic volcanism in Southern Patagonia (47° 15``S), Argentina. Gondwana 11 Conference. p 45 Christchurch

Sruoga, P., Japas, S., Salani, F. y Kleiman, L., 2008. La Peligrosa Caldera: A piecemeal transtensional structure in the Jurassic Chon Aike silicic LIP of Southern Patagonia (47° 15´S), Argentina. IAVCEI General Assembly

Two-phase, reciprocal, double trapdoor collapse at Hannegan caldera, North Cascades, Washington, USA

David S. Tucker
Mount Baker Volcano Research Center
Geology Department
Western Washington University
516 High Street
Bellingham, Washington 98225-9080 USA
DaveTucker@mbvo.wwu.edu

Keywords: Hannegan caldera, trapdoor collapse, welded tuff, ignimbrite, collapse mechanism

1. INTRODUCTION

The deeply dissected, well exposed Pliocene-aged Hannegan caldera (3 x 8 km) in the northern Cascade arc (Fig. 1; Tucker, 2006; Tucker *et al.,* 2007) encloses two 900-m-thick high-silica rhyolite ignimbrites (74-79 wt % SiO_2). The ignimbrites are bounded by a single elliptical fault. Collapse occurred in two events, each most likely a trapdoor, with the second reciprocal in direction to the first. The initial collapse has been dated by $^{40}Ar/^{39}Ar$ to 3.722 ± 0.020 Ma (Hildreth *et al.*, 2003). The caldera is located in rugged mountain wilderness of the North Cascades range, and extends from temperate conifer forests at 915 m a.s.l to glaciated alpine peaks over 2100 m a.s.l. Hannegan caldera is the oldest surviving component of an age-progressive alignment of magmatic centers that extend southwest 25 km to the Holocene Mount Baker volcano (Hildreth *et al.,* 2003). The 1.15 Ma Kulshan caldera (Hildreth, 1996) lies 12 km southwest of the Hannegan caldera along the magmatic alignment.

2. INTRACALDERA ROCKS

The rocks filling the caldera are the Hannegan volcanics (Misch, 1966; Staatz *et al.*, 1972; Tabor *et al.,* 2003). These rocks are described and interpreted in a caldera context in Tucker *et al.,* (2007); a color geologic map and cross section can be found on the Internet (Tucker, 2006). Two lithic-rich ignimbrites dominate the assemblage, each > 900 m thick; the older, northern, densely welded *ignimbrite of Hannegan Peak* (Figure 2) underlies the southern *ignimbrite of Ruth Mountain.* Precaldera volcanism is preserved only as andesitic breccia within the older ignimbrite. Wall rock breccia lenses and megaclasts are intercalated in both ignimbrite units. A layer of lacustrine sedimentary rocks, as much as 30 m thick, separate the ignimbrites at some localities. The ignimbrites are intruded by hundreds of dikes with compositions ranging from basaltic andesite (54.7 wt % SiO_2) to rhyolite (73% wt % SiO_2). Two plutons intrude the caldera's southwest corner, with U-Pb ages of 3.42 ± 0.10 and 3.36 ± 0.20 Ma (Tucker *et al.*, 2007). A lava erupted at 2.96 Ma.

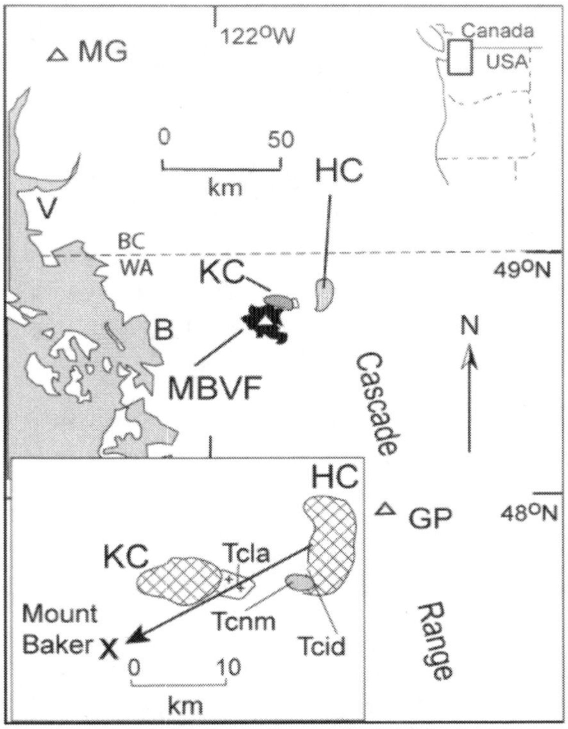

Figure 1: Location of Hannegan caldera (HC) in Washington State, USA. MBVF: Quaternary Mount Baker volcanic field. KC- 1.15 Ma Kulshan caldera; V- Vancouver, British Columbia; B- Bellingham, Washington. Inset: Heavy line shows southwestward migration of magmatic focus from HC to MBVF. Tcnm, Tcid: post-Hannegan plutons; LA: Lake Ann stock. Modified from Tucker *et al.* (2007).

Figure 2: *Ignimbrite of Hannegan Peak* on the south face of Hannegan Peak, 1886 m. The face shown here is ~330 m in height. Pale gray, nonwelded, vapor phase ignimbrite forms the summit cap. Densely welded tuff forms the rest of the face, continuing for hundreds of meters below the bottom of the photo. A lens of wall rock breccia forms the dark band at left center. The shape of the apparently curved felsite dike that cuts the tuff at right center results from gully-and-rib topography. At the lower right, 1-5 m stratified ignimbrite bands dip shallowly into the caldera, representing pulsed accumulation during the first phase of eruption and collapse.

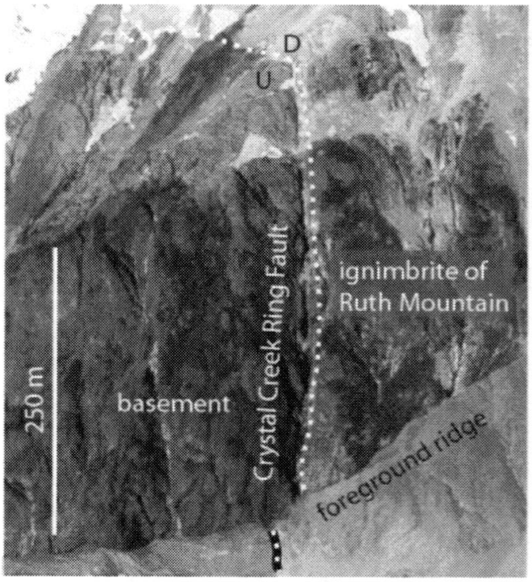

Figure 3: Aerial view looking west to the ring fault on the south margin of Hannegan caldera. The fault forms the vertical gully at center. The visible portion of the dike-intruded fault is 250 m high. A foreground ridge obscures another 120 m of

the fault; the fault continues on the east side of this ridge (dashed line). Schist wall rock (left) is juxtaposed at the fault against ignimbrite of Ruth Mountain. Modified from Tucker (2004).

3. EVIDENCE FOR RECIPROCAL DOUBLE TRAP DOOR COLLAPSE

The entire caldera is enclosed within a single spectacularly exposed, vertically dipping, ring fault (Figure 3). Analogue modeling of caldera collapse that generate elongated calderas commonly begin with trapdoor collapse (*e.g.* Kennedy *et al.*, 2004; Acocella, 2007, and references therein). The densely welded *ignimbrite of Hannegan Peak* is exposed only in the northern half of the caldera; foliations dip generally southward. About half way across the caldera the *ignimbrite of Ruth Mountain* appears and covers the older ignimbrite. The east-west contact runs completely across the caldera at *ca.* 1300 m a.s.l., dipping about 35 degrees south (Figure 4; Figure 5). Both intracaldera ignimbrites appear on high summits within the caldera: the 900-m-thick northern ignimbrite at the summit of Hannegan Peak (1886 m) and the southern unit atop Ruth Mountain (2169 m), 3.5 km south of Hannegan Peak. In this interpretation (Figure 6), the enclosing ring fault formed in two phases, first in the north, later in the south, by sequential trapdoor collapses. See Tucker *et al.* (2007) for a full treatment.

Figure 4: View looking east of south-dipping contact between ignimbrite of Hannegan Peak and overlying ignimbrite of Ruth Mountain, *ca*. 100 m distant. Elsewhere the units are separated by south dipping lacustrine mudstones. Steepness of dip is exaggerated in downward oblique view. Dipping contact results at least in part from down-to-the south trapdoor rotation during second phase of Hannegan caldera collapse.

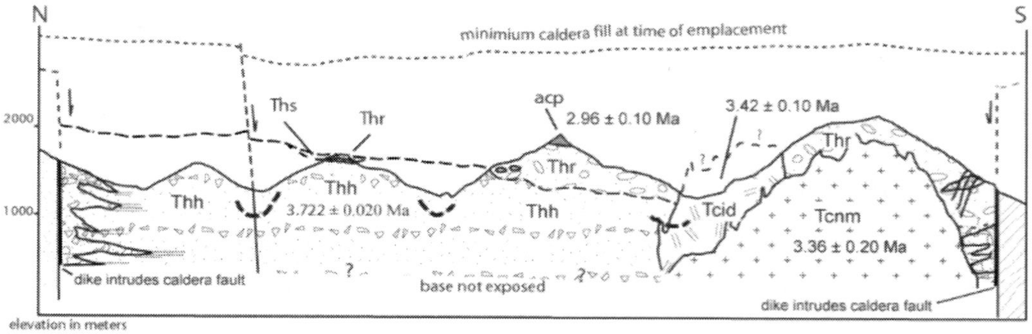

Figure 5: North-south cross section through Hannegan caldera. Units: Thh- ignimbrite of Hannegan Peak; Thr- ignimbrite of Ruth Mountain; Ths- intraignimbrite shale; acp- post caldera lava; Tcid and Tcnm- postcaldera plutons; acp- andesite lava. Heavy dashed lines are deepest exposures in valleys. Contact between intracaldera ignimbrites dips southward. Note breccia lenses inboard of the vertical ring fault. At least 1 km of erosion has occurred since Tcnm was intruded.

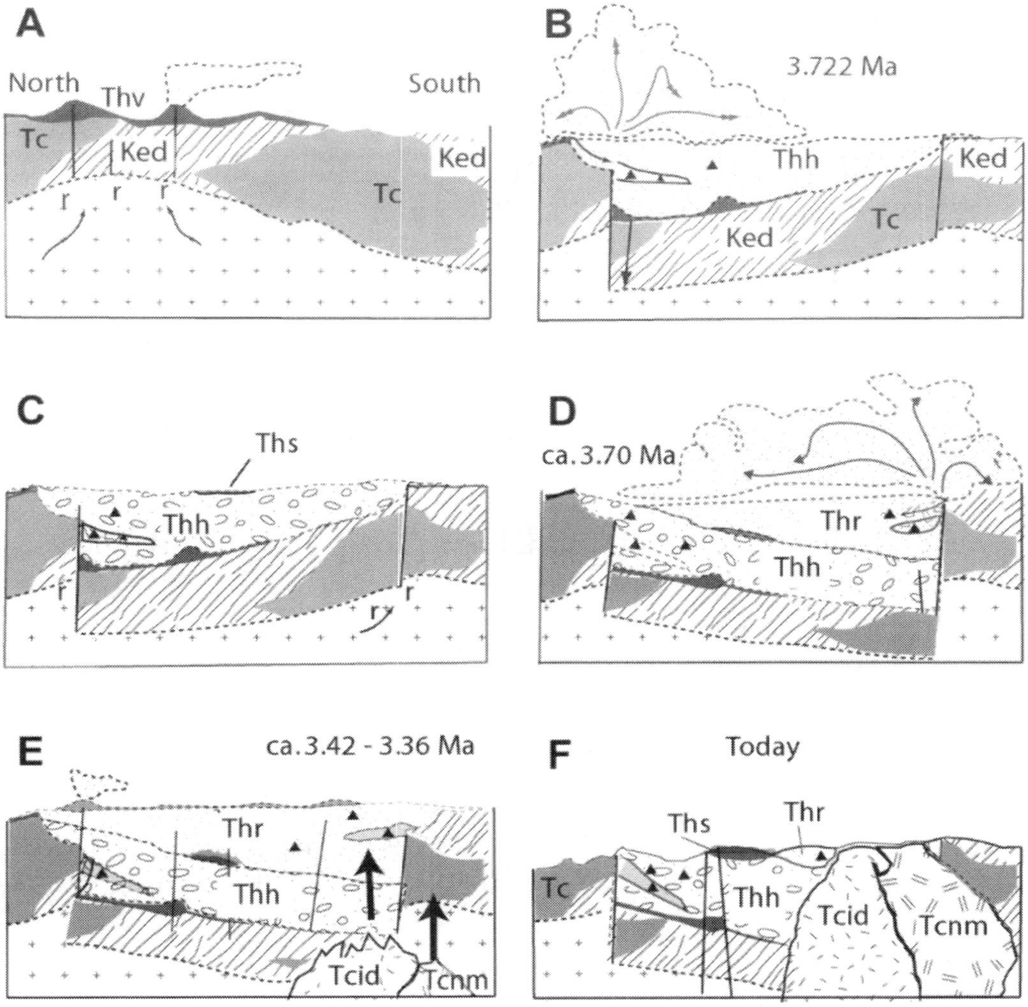

Figure 6: Schematic illustration of Hannegan caldera development. Precaldera topography is hypothetical. Units and symbols as in Figure 5, plus: Thv- precaldera volcanics; Ked- schist basement; Tc- batholith basement; *r*- volatilized rhyolite magma; closed triangles- wall rock megabreccia.

A) High silica rhyolite magma differentiates at the top of a magma chamber with asymmetrical upper surface. **B)** Voluminous eruption of Hannegan Peak ignimbrite, with down-to-the-north trapdoor caldera collapse. A hinge or small-displacement fault borders the southern margin of the caldera. **C)** Period of sedimentation in localized lakes. Volatilized rhyolite magma continues to differentiate and accumulate in cupola resulting from initial trap-door collapse. **D)** Ignimbrite of Ruth Mountain erupts during second-phase collapse, with intercalated breccias. First phase trap-door block rotates downward to the south as the ring fault propagates around the southern portion of the caldera. **E)** Devolatilized remnant magma is intruded into ring faults. Less evolved magmas intrude as dikes. Two plutons intrude caldera's southwest margin. **F)** Modern erosional surface develops, after removal of at least 1 km of caldera fill, exposing the plutons on summits. Modified from Tucker *et al.*, 2007.

REFERENCES

Acocella V., 2007, Understanding caldera structure and development: An overview of analogue models compared to natural calderas: Earth-Science Reviews 85(3-4), 25-160. doi:10.1016/j.earscirev.2007.08.004.

Hildreth, W., 1996, Kulshan Caldera: a Quaternary subglacial caldera in the North Cascades, Washington: Geological Society of America Bulletin, 108(7), 786-793. doi:10.1130/0016-7606(1996)108<0786:KCAQSC>2.3.CO;2.

Hildreth, W., Fierstein, J., Lanphere, M., 2003, Eruptive history and geochronology of the Mount Baker volcanic field, Washington: Geological Society of America Bulletin, v. 115, 729-764. doi:10.1130/0016-7606(2003)115<0729

Kennedy, B., Stix, J., Vallance, J., Lavalleé, Y., Longpré, M-A., 2004, Controls on caldera structure: Results from analogue sandbox modeling: Geological Society of America Bulletin, 116(5-6), 515-524. doi: 10.1130/B25228.1

Misch, P., 1966, Tectonic evolution of the Northern Cascades of Washington State; a west-cordilleran case history: Special Volume – Canadian Institute of Mining and Metallurgy, v. 8, 101-148.

Staatz, M.H., Tabor, R.W., Weis, P.L., Robertson, J.F., Van Noy, R.M., Pattee, E.C., 1972, Geology and mineral resources of the northern part of the North Cascades National Park, Washington: U.S. Geological Survey Bulletin, Report: B 1359, 132 p.

Tabor, R.W., Haugerud, R.A., Hildreth, W., Brown, E.H., 2003, Geologic map of the Mount Baker 30 X 60 minute quadrangle, Washington: U.S. Geological Survey Map I-2660, scale 1:100,000, 2 sheets.

Tucker, D.S., 2004, Geology and eruptive history of Hannegan caldera, North Cascades, Washington: Bellingham, Western Washington University, USA, M.Sc. thesis, 125 p.

Tucker, D.S., 2006, Geologic Map of the 3.72 Ma Hannegan caldera and associated rocks, North Cascades, Washington, scale 1:24,000: Geological Society of America Digital Map and Chart Series 3, 2 sheets, , doi: 10.1130/2006.DMCH003.

Tucker, D. S., Hildreth, W., Ullrich, T., Friedman, R., Geology and complex collapse mechanisms of the 3.72 Ma Hannegan Caldera, North Cascades, Washington, USA ;Geological Society of America Bulletin, 119(3-4):329-342 doi: 10.1130/B25904.1

Hydro-geochemical and isotopic fluid evolution of the Los Azufres caldera geothermal field, central Mexico

González-Partida, E. [1], **Viggiano-Guerra, J.C.**[2], **Pérez, R.J.** [3]
[1]Centro de Geociencias, Campus Juriquilla-UNAM, Querétaro, México, egp@geociencias.unam.mx

[2]Subgerencia de Estudios C.F.E., Morelia. Michocán, México, cesar.viggiano@cfe.gob.mx

[3]Universidad de Calgary, Canada, rene@geochemicalengineering.com

Keywords: geothermal reservoir, fluid inclusions, melting temperatures, boiling.

Summary

Hydrothermal alteration at Los Azufres geothermal field is mostly propylitic showing progressive dehydration and temperature increase with depth. Regular argillic and advanced argillic zones overlie the propylitic zone due to intense gas activity of the hydrothermal system. The deepest fluid inclusions (proto-fluid) are liquid-rich and with low salinity, with NaCl as the dominant fluid type and ice melting temperatures (Tmi) near 0 °C, and salinities of 0.8 wt% NaCl-equivalence. The homogenization temperature (Th) is 335±5 °C. The boiling zone has a Th of ± 300 °C and apparent salinities between 1 and 4.9 wt% NaCl-equivalence, implying a vaporization process and a very important participation of non-condensable gases (NCGs), mostly CO_2. We interpret the evolution of this system as deep liquid water that boils during ascension through fractures connected to the surface. Boiling is caused by a drop of pressure, which favors an increase in the steam phase within the brine ascending towards the surface. During the ascent, the fluid becomes steam-dominant in the shallowest zone and mixes with meteoric water in perched aquifers. Stable isotope compositions ($\delta^{18}O$ – δD) of the geothermal brine indicate mixing between meteoric water and a minor magmatic component. The enrichment in $\delta^{18}O$ is due to the rock–water interaction at relatively high temperatures.

IOP Conf. Series: Earth and Environmental Science **3** (2008) 012012 doi:10.1088/1755-1307/3/1/012012

1. Introduction

The geothermal reservoir of Los Azufres is located in the eastern central portion of the Michoacan State, central Mexico and within the Trans-Mexican Volcanic belt (Fig. 1). It was discovered in 1972, and actually, the geothermal field has an installed capacity of 88 MW in an area of 35 km^2. There are 67 drilled wells, from which 37 are productive, with an average production of 39 t/h of steam. The depth of the productive wells ranges from 627 to 3544 m. The 88 MW are generated with 7 units of 5 MW each, a condensation unit of 50 MW, and two units of binary cycle of 1.5 MW each. There are plans to expand the capacity of the geothermal field to 128 MW with the installation of two more units of 20 MW.

2. Surface geology

The local basement includes a sequence of andesitic and basaltic lavas and interbedded volcanic agglomerates and paleosoils. Unconformably overlying this sequence are silicic volcanic rocks (Agua Fría rhyolite), lake sediments, dacitic lavas (San Andrés dacite), ignimbrites and domes (Yerbabuena rhyolite). The last volcanic stage at Los Azufres is represented by basaltic lavas and pyroclastic deposits and cinder cones in the east and west of the geothermal field.

IOP Conf. Series: Earth and Environmental Science 3 (2008) 012012 doi:10.1088/1755-1307/3/1/012012

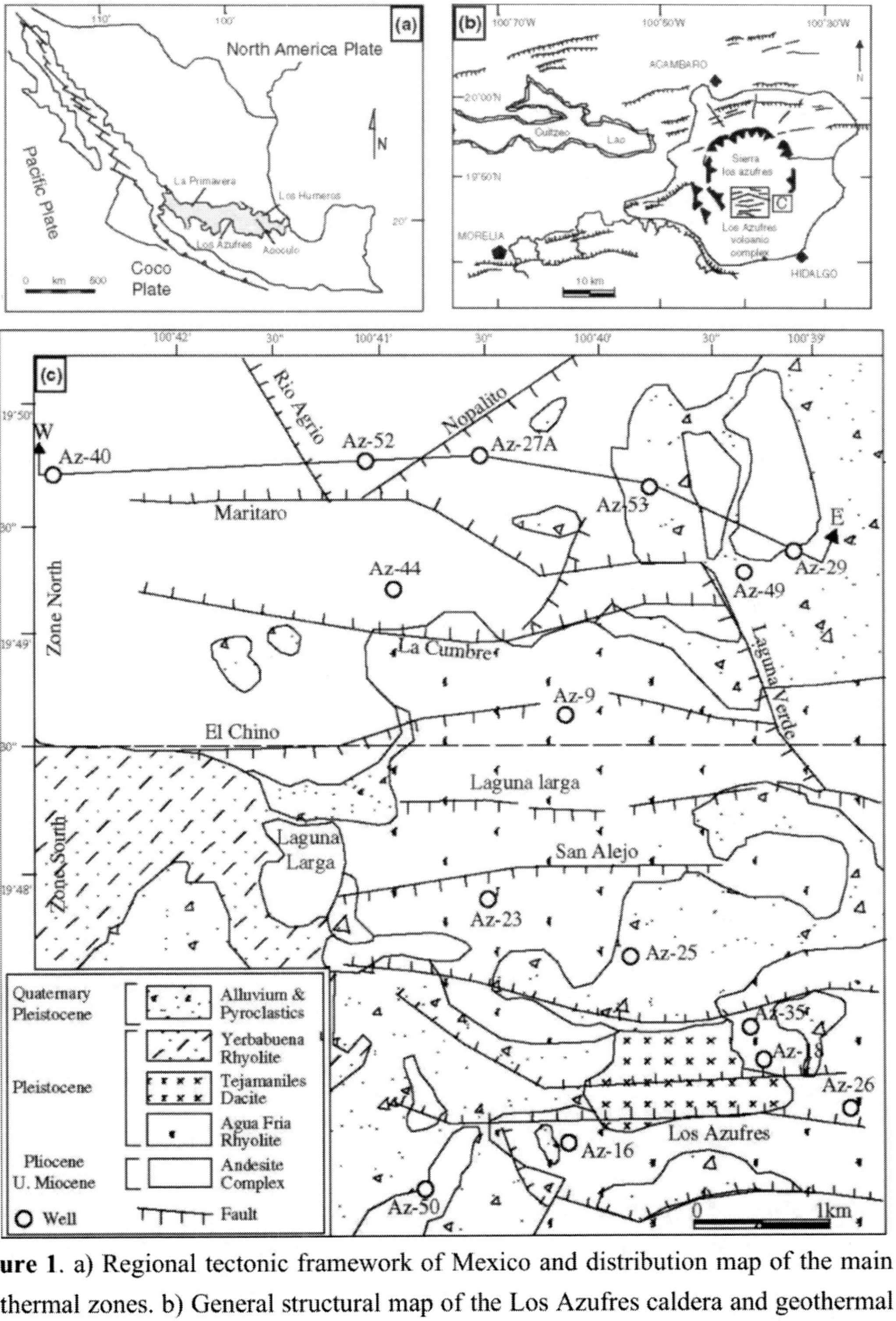

Figure 1. a) Regional tectonic framework of Mexico and distribution map of the main geothermal zones. b) General structural map of the Los Azufres caldera and geothermal field. c) Detailed geologic map of Los Azufres, location of the geothermal wells and location of the cross-section W–E (wells Az-40, Az-52, Az-27A, Az-53 and Az-29). Modified from González-Partida et al. (2000 a, 2000 b).

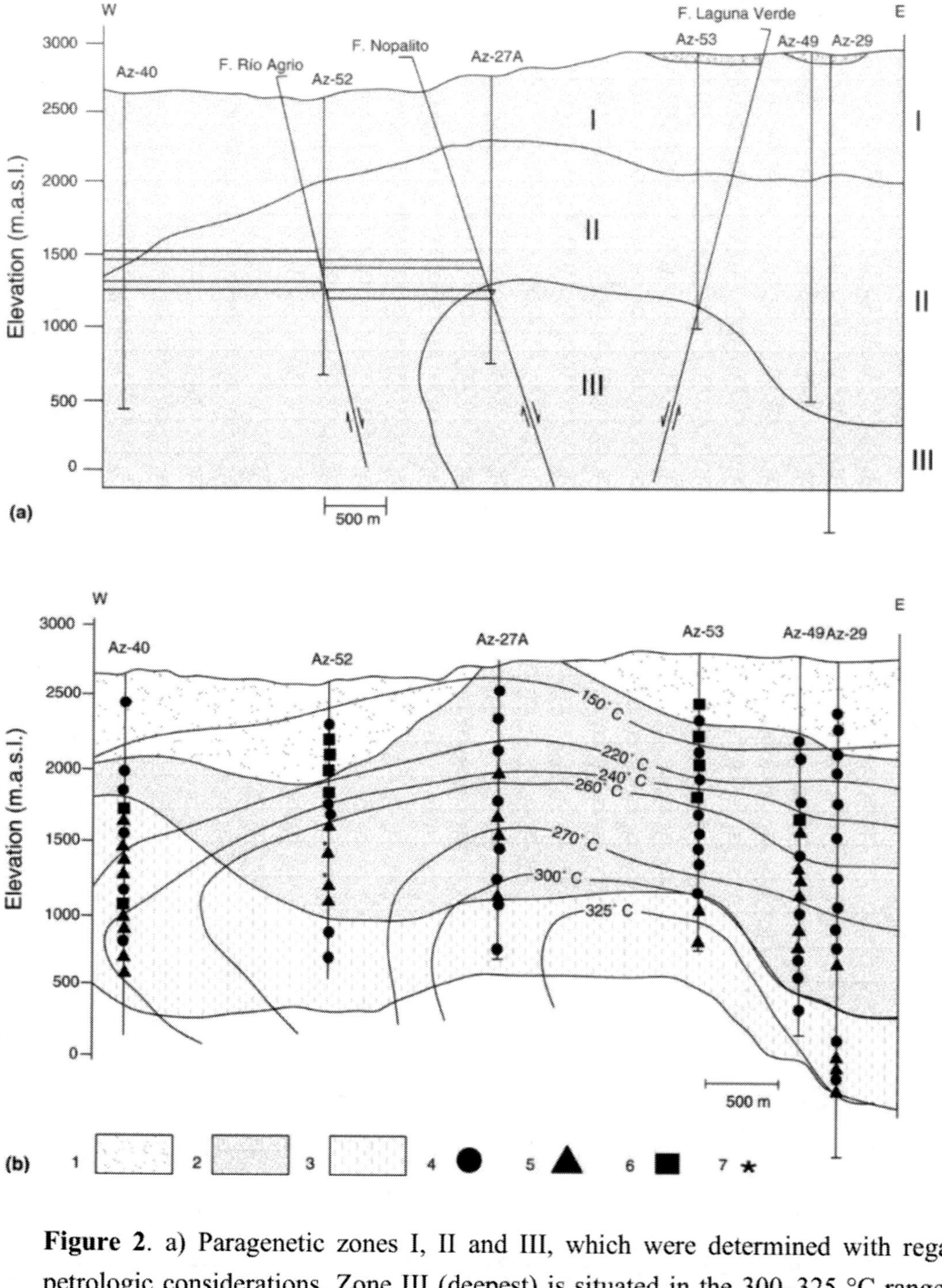

Figure 2. a) Paragenetic zones I, II and III, which were determined with regard to petrologic considerations. Zone III (deepest) is situated in the 300–325 °C range; it is characterized by the assemblage epidote + clinozoisite + actinolite + tremolite ± garnet and smectite+illite+chlorite in the clay fraction (more chlorite than illite and smectite). Zone II is situated in the 220–300 °C range, with the mineral assemblage

epidote+laumontite wairakite, and within the clay fraction illite+ smectite+chlorite. Zone I is situated in the 150–220 °C range and is defined by the predominance of laumontite. b) W–E cross-section indicating isoconcentration zones of the paleo-brine (fluid inclusions) and isotherms obtained from microthermometric analyses of fluid inclusions. Continued lines correspond to average homogenization temperatures zones, where (1) clathrate positive fusion temperaturas, (2) fusion temperatures from 0.7 to 2.4 °C, (3) fusion temperatures from 0.5 to 0.7 C, (4) microthermometric analysis of quartz (d), (5) microthermometric analysis of epidote (m), (6) microthermometric analysis of calcite (j), (7) microthermometric analysis of prehenite (*). Modified from González-Partida et al. (2000 a, 2000 b).

3. Hydrothermal alteration.

A generalized zoning of the calc-silicate assemblage can be observed in the alteration zones, with zeolites in the upper part and epidote–clinozoisite in the deepest levels (González-Partida, 2000). An argillic alteration zone overlies the calc-silicate zone and is the dominant surface manifestation of the hydrothermal alteration. In some parts, there is a mineral assemblage composed of kaolinite–alunite–S–quartz (advanced argillic zone) formed by the interaction of steam and shallow groundwater.

4. Geochemistry of present day fluids

Los Azufres geothermal area is dominated by Cl and Na. The Cl concentration ranges between 1485 and 7297 mg/L, and Na concentration varies between 914 and 4442 mg/L. The next highest dissolved species are SiO_2 and K. Boron concentration is relatively high but the Ca and Mg concentrations are relatively low (González-Partida et al., 2000 a, 2000 b). The water from all the geothermal wells is ''mature'' water type (Cl-rich), which is characteristic of deep geothermal water. Water from the hot springs, is heated by geothermal steam (SO_4 rich). Water from the periphery is mainly meteoric water with little or no geothermal influence. Samples from hot springs fall in the Cl-rich type suggesting that this geothermal water reaches the surface through faults and fractures. The acid SO_4 - rich water is meteoric water that interacts with geothermal steam. The SO_4-rich water is present in geothermal and volcanic systems where steam condensate interacts with shallow meteoric water causing oxidation of H_2S to SO_4. The high concentrations of B in this type of water (B = 0.08 and 7.97 mg/L) and the relatively low concentrations of Mg (Mg = 0.03 and 112.5 mg/L) suggest a deep circulation. Furthermore, the water/rock interaction must have taken place at relatively high temperature because chlorite solubility decreases with increase in temperature, thus lowering the concentration of Mg. All other waters from the springs fit a HCO_3-Cl and HCO_3 Cl-poor trend (<12 mg/L), with a low B concentration (<1 mg/L). However, the Ca and Mg concentrations of these intermediate waters are high, suggesting that they are meteoric waters with shallow circulation through volcanic rocks. The gases in the steam dominant zone are mostly composed of CO_2 (80–98% of all the gases). Other gases present are: H_2S (1.16%), H_2 (0.026%), N_2 (0.77%), and CH_4 (0.01%). Well samples show a positive $\delta^{18}O$ compared to the meteoric line, which is a characteristic of geothermal fluids. This enrichment in $\delta^{18}O$ ‰ is due to water–rock interaction at relatively high temperatures (Arnold and González-Partida, 1987, Birkle et al., 1997). Most of the spring samples plot close to the global meteoric water line.

5. Fluid inclusions composition

At Los Azufres, fluid trapped in inclusions during the growth of hydrothermal minerals in veins (calcite, prehnite, wairakite, anhydrite, quartz and epidote) give a record of the earliest known fluids in this active hydrothermal system.

The iso-concentrations of the proto-fluid and the average Th are indicated in Fig. 2, which is a cross-section based on 6 geothermal wells. The fluid inclusions have an initial ice melting temperature (eutectic temperature, Te), Te= 21.5 ± 0.5 °C, which implies a NaCl rich brine. The final ice melting temperatures (Tmi) reveals a zoning distribution: (a) deepest zone, close to the base line 0 (zone 3 in Fig. 2), where Tf = 0.5 to 0.7 °C, with salinities between 0.88 and 1.23 wt% equivalent NaCl; (b) intermediate zone, located between 1000 and 2500 m.a.s.l. (zone 2 in Fig. 2), with a mixture of L + V and V fluid inclusions, and Tmi = 0.7 to 2.4 °C, which corresponds to apparent salinities of ap.1.2–4.9 wt% NaCl; and (c) shallowest zone located between the surface and the steam-rich zone (steam cap or ''clathrate cap'' zone 1 in Fig. 2), with positive values for the fusion temperature, indicating a strong chemical change in the paleo-brine (proto-fluid). The fusion temperatures in this zone are Tmi = +0.1 to +7 °C, which implies an invasion of CO_2 with some CH_4,N_2, CO, H_2 and H_2S. The measured homogenization temperatures, plotted as isotherms (Fig. 2), show an ascending zone between wells Az-29 and Az-27A, and a descending zone in well Az-40 (this well has a thermal inversion at depth). Boiling processes start at the Th = ±300 °C isotherm, which coincides with an increase in the apparent salinity determined, and an increase of gas concentrations in the fluid.

References

Arnold, M. and González-Partida, E., 1987. Equilibrio Químico e Isotópico Entre Sulfatos y Sulfuros de los fluidos hidrotermales del Campo Geotérmico de Los Azufres Mich.; Implicaciones sobre el origen del azufre. *Revista del Instituto de Geología*, 7, 1-20.

Birkle, P., Torres, R., González-Partida, E., 1997. Escurrimiento Superficial en los Acuíferos del Campo Geotérmico de Los Azufres Mich. *Ingeniería Hidráulica en México*, XII/3-2, 37-48.

González-Partida, E., Birkle, P. and Torres-Alvarado, R., 2000a. Evolution of the hydrothermal system at the geothermal field of Los Azufres México, based on fluid inclusions, isotopic and petrologic. *Journal of Volcanology and Geothermal Research,* 104, 277-296.

González-Partida, E., Tello Hinojosa, E. and& Verma, P., 2000b. Análisis geoquímico e isotópico de aguas geotérmicas y manantiales para definir el estado de equilibrio agua-roca del reservorio de los Azufres, Mich., México. *Ingeniería Hidráulica en México,* Vol. XV/3, 89-99.

González-Partida, E., 2000. Evidencias de evolución de un fluido básico a ácido a partir del análisis de la alteración hidrotermal del campo geotérmico de Los Azufres Michoacan: Su relación con los yacimientos de baja sulfuración fósiles. *Revista Mexicana de Ciencias Geológicas,* 17/1, 76-82.

Pre-eruptive conditions of the phonolitic magma from the El Abrigo caldera-forming eruption (Las Cañadas caldera, Tenerife, Canary Islands)

Martí, J[a]., Andújar, J. [a], Costa, F.[a], Wolff, J.A.[b] and Carroll, M.R.[c]

a. Institute of Earth Sciences "Jaume Almera", CSIC, C/ Lluis Solé I Sabaris, s/n Barcelona, 08028 Spain.

b. School of Earth and Environmental Sciences, Washington State University, Pullman, WA 99164-2812, USA. E-mail address: jawolff@mail.wsu.edu

c. Dipartimento di Scienze della Terra, Via Gentile III da Varano, Università di Camerino, 62032 MC, Italy. E-mail ardes: Michael.carroll@unicam.it

Keywords: phase equilibria, ignimbrite, caldera, experimental petrology, Las Cañadas caldera, phonolite, Tenerife.

1. Introduction

Precise knowledge of intensive variables for caldera related magmas (the so-called pre-eruptive conditions) is essential to improving our understanding of these catastrophic eruptions. One of the most precise methodologies for determining the temperature (T), pressure (P), volatile composition and content of the magma is to perform phase equilibrium experiments (e.g., Hammer and Rutherford, 2003; Pichavant et al., 2007) where the phase assemblages and compositions of the volcanic rocks are experimentally reproduced. This approach has been successfully applied to a number of volcanoes including Mount St. Helens, (e.g., Rutherford et al., 1985; Gardner et al., 1995), El Chichón, (Luhr, 1990), Novarupta, (Hammer et al., 2002), Mt. Pelée (Martel et al., 1998; Pichavant et al., 2002), Soufriere Hills (Barclay et al., 1998), Unzen (Sato et al., 1999) and Volcan San Pedro (Costa et al., 2004). Fewer studies have been made related to collapse calderas (e.g., Pinatubo, Rutherford and Devine, 1996; Scaillet and Evans, 1999; La Garita, Johnson and Rutherford, 1989; Santorini, Cottrell et al., 1999; Laacher See, Berndt et al., 2001; Harms et al., 2004).

2. The Las Cañadas caldera

The Las Cañadas caldera in Tenerife (Fig. 1) is the result of vertical collapses related to explosive eruptions (Martí et al., 1994, 1997; Martí and Gudmundsson, 2000). Within the Las Cañadas edifice Martí et al. (1994) identified a Lower Group (LG) representing a complex constructive phase, and an Upper Group (UG, Fig. 2). The latter is composed of phonolitic pyroclastic deposits and subordinate mafic rocks, and has been subdivided into three major formations each associated with a cycle of phonolitic explosive eruptions with vents in the area now occupied by the caldera (Martí et al., 1994, 1997): (1) Ucanca (1.57-1.07 Ma), (2) Guajara (0.85-0.57 Ma) and (3) Diego Hernández (>0.37-0.19 Ma). Each cycle culminated in caldera collapse, and was followed by migration of the focus of volcanic activity to the northeast, resulting in the progressive enlargement of the Las Cañadas caldera complex (Martí et al., 1994; Martí and Gudmundsson, 2000). The uppermost unit of the Diego Hernández Formation (Edgar et al., 2007) and the most widespread deposit of the Las Cañadas Upper Group is the Abrigo ignimbrite, a caldera-forming deposit that covered the whole island of Tenerife (with the exception of the Anaga massif, Fig.1) and which is the focus of this study.

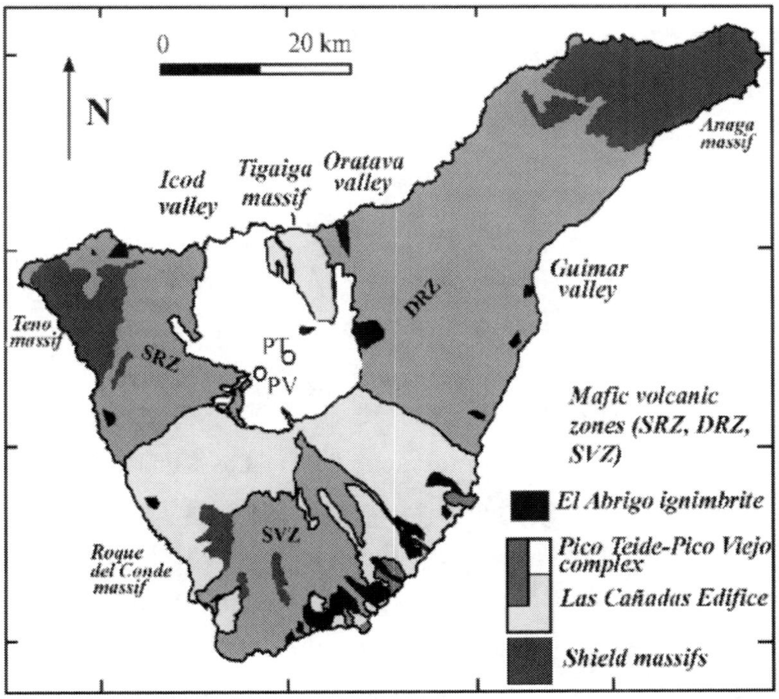

Figure 1: Simplified geological map of Tenerife illustrating the outcrops of the products from the basaltic shield, Las Cañadas edifice, the active volcanic complex of Teide and Pico Viejo, and the distribution of the Abrigo ignimbrite on Tenerife. PV: Pico Viejo, PT: Teide. Post-shield mafic volcanic zones are SRZ: Santiago del Teide Rift/Ridge Zone; DRZ: Dorsal Rift/Ridge Zone; SVZ: Southern Volcanic Zone. Figure modified from Ablay and Martí (2000).

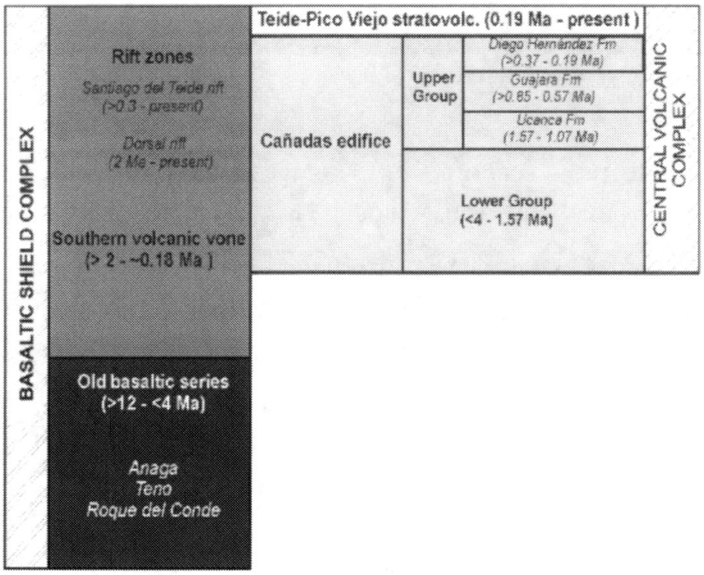

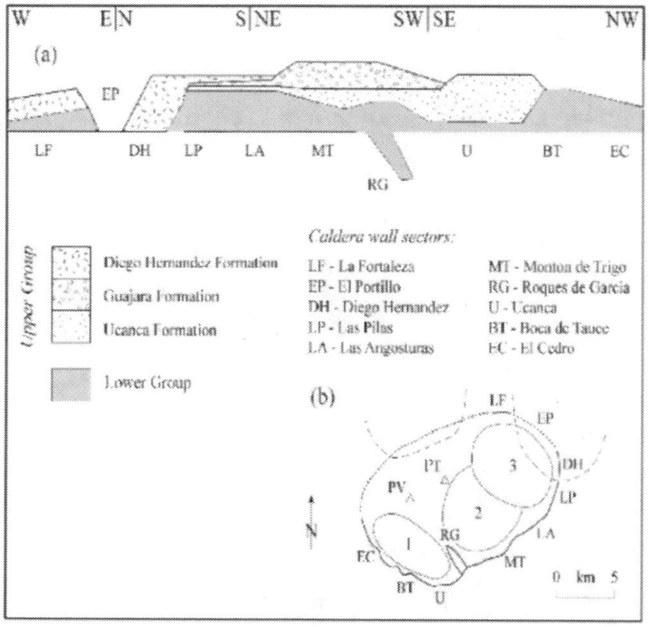

Figure 2. a) Schematic illustration of the stratigraphy of the Las Cañadas caldera wall, showing the sequence of the Lower Group and three formations of the Upper Group (From Edgar, 2003, after Martí et al., 1994). b) Map showing the caldera wall sectors and outlines of the three major vertical collapse calderas as proposed by Martí et al. (1994): 1 - Ucanca, 2 - Guajara, 3 - Diego Hernández (i.e. younging to the NE). The inferred positions of the La Orotava (LO) and Icod (I) landslide scarps are also shown (dashed lines). PG: Pico de Guajara, PT: Teide, PV: Pico Viejo

IOP Conf. Series: Earth and Environmental Science **3** (2008) 012013 doi:10.1088/1755-1307/3/1/012013

3. Petrological investigations

We have performed phase equilibrium experiments to determine the pre-eruptive conditions of the El Abrigo ignimbrite (~20 km^3 of DRE). The Abrigo phonolite contains low amounts of phenocrysts (ca. 7 vol %), mainly sanidine, biotite, clinopyroxene, and magnetite, with minor amounts of titanite, apatite, and haüyne. We have investigated pressures from 50 to 250 MPa, temperatures of 700-900°C, water contents in the melt from 1.5 to ca. 10 wt %, and oxygen fugacity (fO_2) from 1 log unit above the Ni-NiO solid buffer to quartz-magnetite-fayalite. Comparison of the natural and experimental phase proportions and compositions indicates that the phonolite at the roof of the Abrigo magma reservoir was at 130 ± 50 MPa (corresponding to ca. 4 - 5 km below the surface), 825 ± 25 °C, with 3 ± 1 wt% dissolved H_2O and fO_2 at the Ni-NiO buffer ± 1 log unit. This shows that the magma that produced the largest ignimbrite on Tenerife was stored at relatively shallow depths but was water-undersaturated, and its eruption was probably triggered by input of fresh mafic magma. Comparison of the Abrigo phase equilibrium results with those obtained in other phonolitic magmas (e.g., Eifel, Kerguelen, Somma-Vesuvius) shows that the major element and probably fluid composition (H_2O, Cl and S) may have a strong effect on the presence or absence of phases and their compositions. Extrapolating phase equilibria results obtained on a specific phonolite to constrain the pre-eruptive conditions of other phonolites should be done with care.

4. Conclusions

The combination of petrological and experimental petrology data suggests that the Las Cañadas edifice developed a phonolitic magmatic system(s) that included several coeval magma chambers. These magma chambers had minimum volumes ranging from 1 to > 20 km^3 (Edgar et al., 2007) and were located at shallow depths (3-5 Km). The dry conditions deduced for the El Abrigo phonolitic magmas indicate that they were undersaturated in water at the time of the eruption. Therefore, an overpressure caused by water oversaturation of the magma due to the crystallisation in a closed system cannot be invoked as the triggering mechanism for that eruption. An external agent, such us intrusion of fresh basaltic magma at the bottom of the magma chamber, was required to trigger these eruptions. This is supported by the presence of magma mixing and mingling in El Abrigo pumices. The lack of pumice fall deposits associated with El Abrigo ignimbrite indicate that the eruption immediately developed into massive proportions. This, together with the presence of multiple vents along a ring fault during the El Abrigo eruptions, as it is suggested by the lithics distribution (Pittari et al., 2008), indicates that caldera onset coincided with the beginning of the eruption, when the magma chamber was still overpressurised, not following the classical model of caldera formation in which a significant decompression of the magma chamber is required before starting the chamber roof subsidence.

References

Ablay, G. J., Martí, J., 2000. Stratigraphy, structure, and volcanic evolution of the Pico Teide-Pico Viejo formation, Tenerife, Canary Islands. Journal of Volcanology and Geothermal Research 103, 175-208.

Andújar, J., 2007. Application of experimental petrology to the characterization of phonolitic magmas from Tenerife, Canary Islands. Ph.D. thesis, Universitat de Barcelona, 183 pp.

Barclay, J., Rutherford, M.J., Carroll, M.R., Murphy, M.D., Devine, J.D., Gardner, J., Sparks, R.S.J., 1998. Experimental phase equilibria constraints on pre-eruptive storage conditions of the Soufriere Hills magma. Geophysical Research Letters 25, 3437-3440.

Berndt, J., Holtz , F., Koepke, J., 2001. Experimental constraints on storage conditions in the chemically zoned phonolitic magma chamber of the Laacher See volcano. Contributions to Mineralogy and Petrology 140, 469-486.

Costa, F., Scaillet, B., Pichavant, M., 2004. Petrological and experimental constraints on the Pre-eruption conditions of Holocene dacite from Volcán San Pedro (36°S, Chilean Andes) and the importance of sulphur in silicic subduction-related magmas. Journal of Petrology 45, 855-881.

Cottrell, E., Gardner, J.E., Rutherford M.J., 1999. Petrologic and experimental evidence for the movement and heating of the pre-eruptive Minoan rhyodacite (Santorini, Greece). Contributions to Mineralogy and Petrology 135, 315-331.

Edgar, C.J., Wolff, J.A, Olin, P.H., Nichols, H.J., Pittari, A., Cas, R.A.F., Reiners, P.W., Spell, T.L., Martí, J., 2007. The late Quaternary Diego Hernandez Formation, Tenerife: Volcanology of a complex cycle of voluminous explosive phonolitic eruptions. Journal of Volcanology and Geothermal Research 160, 59–85.

Gardner, J.E., Carey, S., Rutherford, M., Sigurdsson, H., 1995. Petrologic diversity of

Mount St. Helens dacites during the last 4,000 years: implications for magma mixing. Contributions to Mineralogy and Petrology 119, 224-238.

Hammer, J.E., Rutherford, M.J., 2003. Petrologic indicators of pre-eruptions magma dynamics. Geology 31, 79-82.

Hammer, J.E., Rutherford, M.J., Hildreth, W., 2002. Magma storage prior to the 1912 eruption at Novarupta, Alaska. Contributions to Mineralogy and Petrology 144, 144-162.

Harms, E., Gardner, J.E., Schmincke, H.U., 2004. Phase equilibria of the lower Laacher See Tephra (East Eifel, Germany): constraints on pre-eruptive storage conditions of a phonolitic magma reservoir. Journal of Volcanology and Geothermal Research 134, 135-148.

Johnson, M., Rutherford, M.J., 1989. Experimentally determined conditions in the Fish Canyon Tuff, Colorado, magma chamber. Journal of Petrology 30, 711-737.

Luhr, J.F., 1990. Experimental phase relations of water-and sulfur-saturated arc magmas and the 1982 eruptions of El Chichon volcano. Journal of Petrology 31, 1071-1114.

Martel, C., Pichavant, M., Bourdier, J.-L., Traineau, H., Holtz, F., Scaillet, B., 1998. Magma storage conditions and control of eruption regime in silicic volcanoes: experimental evidence from Mt. Pelée. Earth and Planetary Science Letters 156, 89-99.

Martí, J., Mitjavila, J., Araña, V., 1994. Stratigraphy, structure and geochronology of the Las Cañadas caldera (Tenerife, Canary Island). Geological Magazine 131, 715-727.

Martí, J., Hurlimann, M., Ablay, G.J., Gudmundsson, A., 1997. Vertical and lateral collapses on Tenerife (Canary Islands) and other volcanic ocean islands. Geology 25, 879-882.

Martí, J., Gudmundsson, A., 2000. The Las Cañadas caldera (Tenerife, Canary Islands): an overlapping collapse caldera generated by magma-chamber migration. Journal of Volcanology and Geothermal Research 103, 161-173.

Pichavant, M., Martel, C., Bourdier, J.-L., Scaillet, B., 2002. Physical conditions, structure and dynamics of a zoned magma chamber: Mt. Pelée (Martinique, Lesser Antilles Arc). Journal of Geophysical Research 107, doi: 10.1029/2001JB000315.

Pichavant, M., Costa, F., Burgisser, A., Scaillet, B., Martel, C., Poussineau, S., 2007. Equilibration scales in silicic to intermediate magmas - Implications for phase equilibrium studies. Journal of Petrology 48, 1955-1972.

Pittari, A., Cas, R. A. F., Wolff, J. A., Nichols, H. J., Martí, J., 2008. The use of lithic clast distributions in pyroclastic deposits to understand pre- and syn-caldera collapse processes; a case study of the Abrigo Ignimbrite, Tenerife, Canary Islands. *In*: Gottsmann, J., Martí, J. (eds) "*Caldera Volcanism: analysis, modelling and response*",

Developments in Volcanology, Elsevier, 97-142

Rutherford, M.J., Sigurdsson, H., Carey, S., Davis, A., 1985. The May 18, 1980, eruption of mount St. Helens 1. melts composition and experimental phase equilibria. Journal of Geophysical Research 90, 2929-2947.

Rutherford, M.J., Devine, J.D., 1996. Pre-eruption pressure-temperature conditions and volatiles in the 1991 dacitic magma of Mount Pinatubo. In C.G. Newhall and R.S. Punongbayan, Eds., Fire and Mud: Eruptions and Lahars of Mount Pinatubo, Philippines, p. 751–766. PHIVOLCS and University of Washington Press, Seattle.

Sato, H., Nakada, S., Fujii, T., Nakamura, M., Suzuki-Kamata, K., 1999. Groundmass pargasite in the 1991-1995 dacite of Unzen volcano: phase stability experiments and volcanological implications. Journal of Volcanology and Geothermal Research 89, 197-212.

Scaillet, B., Evans, B.W., 1999. The 15 June 1991 eruption of Mount Pinatubo. I. phase equilibria and Pre-eruption P-T-fO_2-aH_2O. Journal of Petrology 40, 381-411.

Thermodynamic state updated of the volcanic caldera and geothermal reservoir of Los Humeros, Puebla, Mexico

Martínez Reyes, José [1], Gonzalez Partida, Eduardo [1], Perez, Renee J. [2]; Tinoco Michel, Jorge A. [1].

[1] *Centro de Geociencias, Universidad Nacional Autónoma de México Campo de Juriquilla, Qro., México, apartado postal 76230.*

[2] Department of Chemical and Petroleum Engineering, University of Calgary, 500 University Drive, Calgary Alberta, Canada, T2N 1N4

Keywords: geothermal reservoir, dominant steam

Abstract

Based on information of enthalpies of the fluids of wells from the geothermal reservoir of Los Humeros, Puebla, México (wells H-012, H-015, H-016, H-017, H-019, H-039), we determined the thermodynamic conditions of the reservoir comparing the values of enthalpies of the fluids of discharge of the wells with the values published in the literature for different thermodynamic state of the fluids.

1. Introduction

The geothermal field of Los Humeros is located in the part oriental central part of Mexico, approximately 25 km to the NW of Perote's city, in the State of Puebla (Fig. 1). It is in the nested caldera complex of Los Humeros-Los Potreros, dated at about 0.5-0.4 Ma (Yañez and García., 1982; Ferriz and Mahood, 1984). This volcanic complex is in the oriental part of the Volcanic Mexican belt (Robin, 1982), near the border with other province, the Alkaline volcanic province of the Gulf of Mexico (Fig. 1 1).

2. Thermodynamic state from the reservoir

Based on updated data from enthalpy of fluids (gas and brine) of of wells H-012, H-015, H-016, H-017, H-019, H-039 from the geothermal reservoir of Los Humeros, we can observe that fluids of the reservoir have enthalpy values greater than 2400 J/g, which correspond to a thermodynamic state steam (Figs. 2 and 3). From this, we deduce that the geothermal reservoir of Los Humeros is in the steam-domain, as has been proposed in the literature (White et al, 1971; Truesdell & White, 1973; Silvester & Pitzer, 1976; Ellis & Mahon, 1977; Arnorsson et al col., 1982; D'Amore et al, 1990; Verma et al, 1998; Prol-Ledesma, 1998; Gonzalez Partida et al, 2001; Arellano et al, 2003; Tello,

2005). Furthermore, this steam domain in the geothermal reservoir can be accentuated by over-exploitation (extraction of brine without sufficient recharge) of the reservoir (Truesdell, 1991; Gonzalez Partida et al., 2001; Tello, 2005).

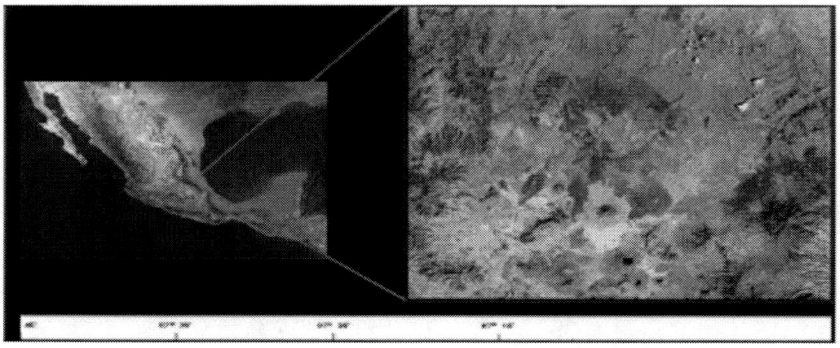

Figure 1: Location of the geothermal area of Los Humeros, Mexico

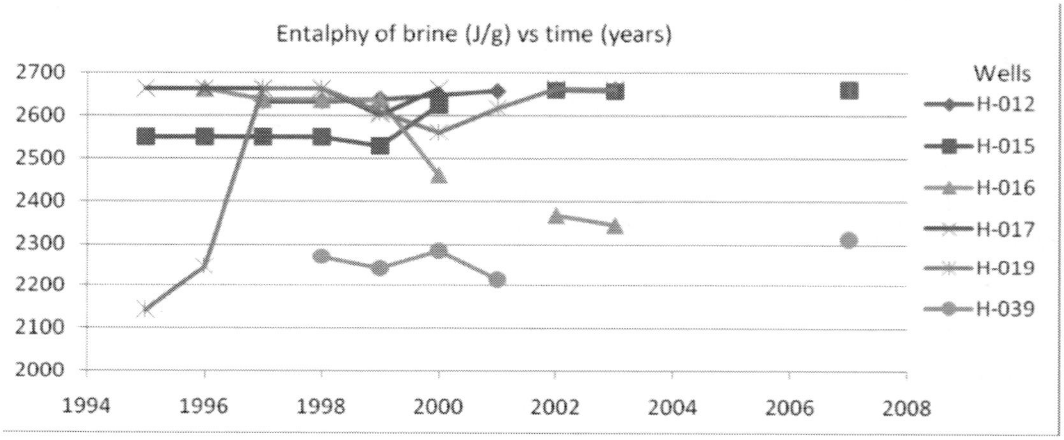

Figure 2: Updated enthalpy values of brine reservoir against time

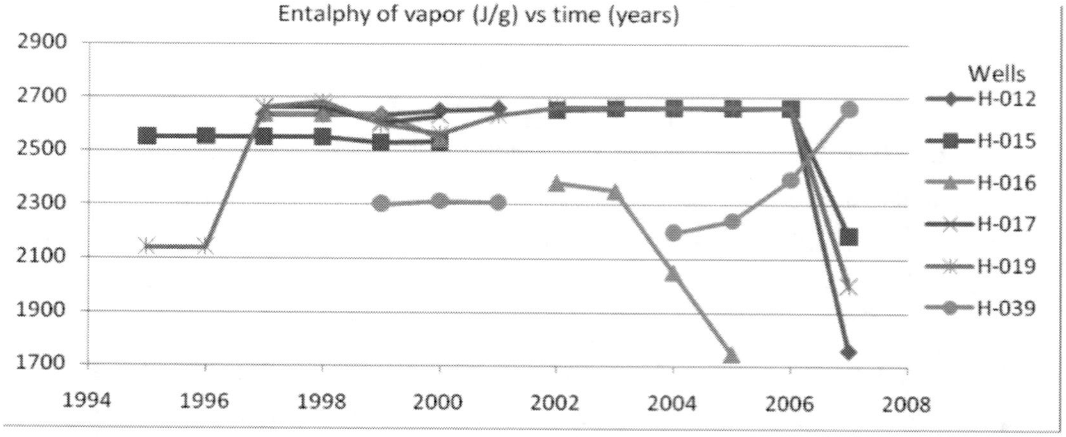

Figure 3: Updated enthalpy values of steam reservoir against time

Acknowledgments

We express our special thanks to the Gerencia de Proyectos Geotermoelectricos de la Comisión Federal de Electricidad by providing us updated data of enthalpy of fluids (brines and gases) from the geothermal wells of Los Humeros geothermal field.

References

Arellano, V. M.; García, A.; Barragán, R. M.; Izquierdo, G.; A. Aragón, A. & Nieva, D., 2003. An updated conceptual model of the Los Humeros geothermal reservoir (Mexico). *Journal of Volcanology and Geothermal Research, Volume 124, Issues 1-2, Pages 67-88*

Arnorsson, S, Gunnlaugsson, E. & Svavarsson, H., 1982. The chemistry of geothermal waters in iceland i. Calculation of aqueous speciation from 0" to 37°C. *Geochimica et Cosmochimica Acta 46,* 1513- *1532.*

Cedillo Rodríguez, F., 1984. Estudio geológico de Los Humeros-Las Derrumbadas, Edos. de Pue. y Ver. Comisión Federal de Electricidad. Informe interna 17/84, 55 p.

C.F.E. (Comisión Federal de Electricidad), 2007. Updated data from enthalpy fluid discharge of a section of wells (H-012, H-015, H-016, H-017, H-019,H-039) of the geothermal reservoir Los humeros, Puebla, Mexico. Not published

D'Amore, F.; Rivera, J.R.; D. Giusti, D. and Rossi, R., 1990. Preliminary geochemical and thermodynamic assessment of the geothermal resources, Sulphur Springs area, St Lucia, W.I. *Applied Geochemistry,* Vol. 5, pp. 587-604

Ellis, A.J. y Mahon, W.A.J., 1977. Chemistry and geothermal systems. Academic Press, New York, 392 p.

Ferriz Dominguez, H. y Mahood, A., 1984. Eruption rates and compositional trends at Los Humeros Volcanic Center, Puebla., Mexico. J. Geophys. Res, V, 89, B10, 8511-8524.

Garduño Monroy, V. H., Romero Ríos, F. y Torres Hernández, R., 1985. Análisis estructural del campo geotérmico de Los Humeros, Pue. Comisión Federal de Electricidad, Informe interno 26/85, 20p.

González Partida, E.; Tello, Hinojosa, E. & Verma, M.P., 2001. Interacción agua geotérmica-manantiales en el campo geotérmico de Los Humeros, Puebla, México. Ingeniería hidráulica en México, vol.:16, núm:2, pàg.:185

Prol-Ledesma, R. M., 1998. Pre- and post-exploitation variations in hydrothermal activity in Los Humeros geothermal field, Mexico. *Journal of Volcanology and Geothermal Research, Volume 83, Issues 3-4, Pages 313-333*

Robin, C. 1982. Relations volcanologie-magmatologie-geodynamique, application au passage entre volcanisme alcalin et andesitique dans le Sud Mexicain (Axe Transmexicain et Province Alcaline Oriental). These de Doc. d, Etat Univ. De Clemont-Ferrand 2503 p.

Silvester, L. F. and Pitzer, K.S., 1976. Thermodynamics of Geothermal Brines. I. Thermodynamic Properties of vapor saturated NaCl (aq). Solutions from 0-300°C. Nat. Tech Informat. Serv. Rep. T1D-4500-R64.

Tello, E., 2005. Estado de equilibrio soluto-mineral y saturacion de minerales de alteracion en fluidos geotermicos de alta temperatura, Tesis Doctoral. Fac. de Ingenieria: Mexico, DF, UNAM- p. 271.

Truesdel A. H., 1991. Origin of acid fluids in geothermal reservoirs. G. R. C. Transactions, V. 15: 289-296.

Truesdell, A.H. and White, D.E., 1973. Production of superheatead steam from vapor-dominated geothermal reservoirs. Geothermics, v. 2, no. 3-4, p. 154-173.

Verma M. P., Tello H. E., Arellano G. V., Nievo G. D., 1998. Acidic fluid in los Humeros geothermal reservoir: A preliminary outlook. Proceedings, Twenty third workshop on Geothermal Reservoir Engineering. Stanford University, Stanford California SGP-TR- 158 26-28.

White, D.E., Mufler, L.J.P. and Truesdell, A.H., 1971. Vapor dominated hydrothermal systems compared with hot-water system, Economic Geology, v.66, no. 1,p.75-97

Yáñez García, C. y Garcia Duran S., 1982. Exploracion de la region geotermica Los Humeros - Las Derrumbadas estados de Puebla y Veracruz. C. F. E., 97 p.

THERMODYNAMIC MODEL FOR THE PHASE EQUILIBRIA OF GASES AND BRINES. EXAMPLE IN THE H_2S–H_2O–NaCl SYSTEM

Martinez Reyes, Jose [1], Perez, Renee J. [2], Heidemann, Robert A. [2], Gonzalez Partida, Eduardo [1], Tinoco-Michel, Jorge A. [1]

[1] Centro de Geociencias, Universidad Nacional Autónoma de México
Campo de Juriquilla, Qro., México, apartado postal 76230 .

[2] Department of Chemical and Petroleum Engineering, University of Calgary,
2500 University Drive, Calgary Alberta, Canada, T2N 1N4

Keywords: Hydrogen sulfide, solubility, water, brine

Abstract
In this work we propose a thermodynamic model for the mixing of gases in water and in aqueous sodium chloride solutions valid to high pressures, high temperatures, and high ionic strength solutions below their critical point. Our model couples Henry's Law with any equation of state to reproduce experimental data in the aqueous-rich liquid and gas-rich vapor region. In our model, the chemical potential of the solute in the brine is related to the chemical potential of the solute in pure water through salting-out coefficients. The model reproduces all crucial phenomena of binary (gas–water) and pseudo-binary (gas–water–salt) vapor–liquid mixtures. We applied the model to reproduce the phase behavior of hydrogen sulfide in NaCl brines. Results and discussions are shown.

1. Introduction
We propose a thermodynamic model for the mixing of gases in water and in aqueous sodium chloride solutions valid to high pressures, high temperatures, and high ionic strength solutions below their critical point. An accurate prediction of the solubility of gases as H_2S in water or in NaCl solutions as a function of pressure, temperature and solution composition over a wide range is important because being the H_2S one of the most plentiful gases in the nature next to the CO_2, is present in natural processes as atmospheric cycles, dissolution of gases in sea waters, geochemical processes as well as in industrial processes (Duan et al., 1996; Perez and Heidemann, 2006, 2007). The chemical literature contains many experimental data on gas solubility in water, but gas solubility data in NaCl brines, at moderate pressures and high temperatures, is scarce.

This is the reason because is important exploring alternative thermodynamic models to reproduce vapor–liquid equilibria data, particularly in NaCl solutions where no experimental data is available (Drumond, 1981; Barta and Bradley, 1985; Carrol and Mather, 1989; Duan et al., 1996).

2. Thermodynamic model

This model couples Henry's Law with any equation of state to reproduce experimental data in the aqueous-rich liquid and gas-rich vapor region respectively (Fernández-Prini and Crovetto, 1989; Fernández-Prini et al., 2003; Perez and Heidemann, 2006, 2007). The components fugacity coefficients of the vapor mixture are calculated with any equation of state (EoS). In this work was used the Soave–Redlich–Kwong SRKEoS (Soave, 1972). For the brine phase, the volatile solutes are modeled using a modified Henry's Law (Harvey 1996; Fernández-Prini and Crovetto, 1989; Fernández-Prini et al., 2003)

In this model, the chemical potential of the solute in the brine is related to the chemical potential of the solute in pure water through salting-out coefficients defined by Setschenov (1889).

3. Example in the H_2S–H_2O–NaCl system

The model was used to reproduce the phase behavior of hydrogen sulfide in NaCl brines. Henry's constants for H_2S in H_2O were obtained directly from Fernandez-Prini et al., (2003) and Harvey (1996). The critical temperature of the pure water is taken as accepted by IAPWS (1995 -647.096ºK). The vapor pressure of the solvent (H_2O) at the temperature of interest is calculated from the correlation of Wagner and Pruss (1993). The critical pressure of the solvent also was taken from IAPWS (1995) (P_{C1}=22064000bars). To calculate the fugacity of the gases in the vapor mixture was used the SRK equation of state of Soave (1972). The hydrogen sulfide partial molar volume at infinite dilution was calculated based on Plyasunov et al. (2000). Experimental data from hydrogen sulfide was extracted from Lee and Mather (1977). Values of hydrogen sulfide's critical pressure, critical temperature, and a centric factor were taken from the literature (Duan et al., 1996) and are 373.4ºK, 89.63 bar and 0.09 respectively.

The H_2S solubilities were calculated with this thermodynamic model under a variety of temperature, pressure and NaCl concentration conditions. The calculated results compared with the experimental data shows that predictions are within experimental uncertainty (Fig. 1).

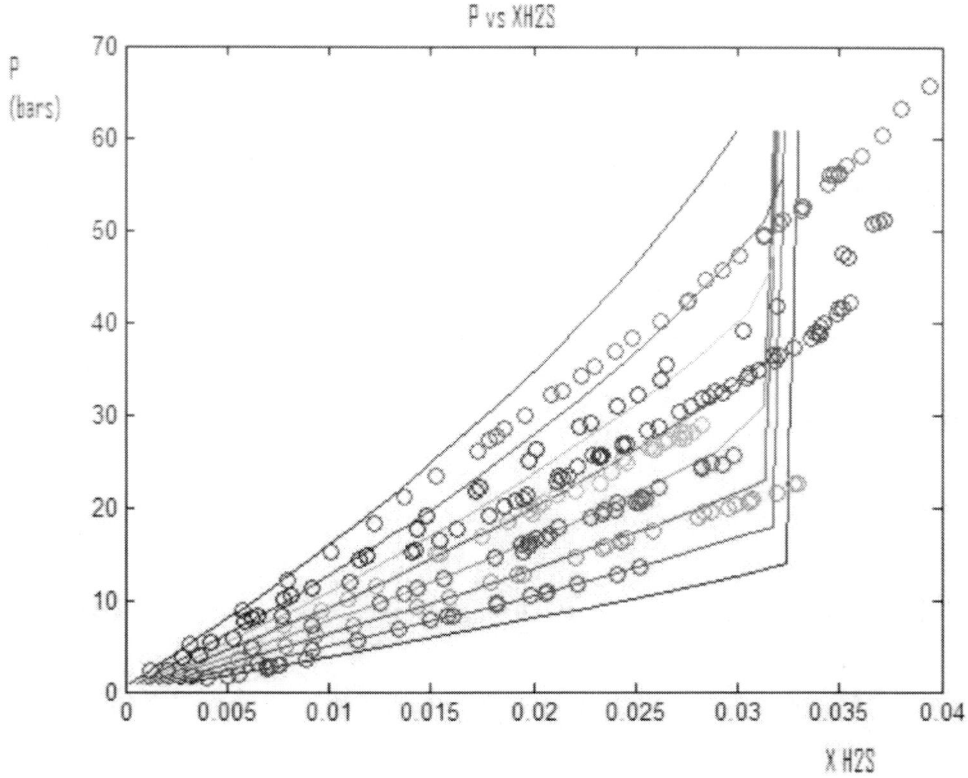

Figure 1. Comparison of experimental H_2S solubility with the predicted model. The solid line is the calculated model and the symbols are data by Lee and Mather (1977).

4. Conclusion

The thermodynamic model presented here predicts internal experimental uncertainty, and the phase behavior of gases in water or brines at the limit of infinite dilution. The model reproduces all crucial phenomena of binary (gas–water) and pseudo-binary (gas–water–salt) vapor–liquid mixtures within and beyond the pressure range of experimental data below their critical point.

References

Barta, L. and Bradley, D.J., 1985. Extension of the specific interaction model to include gas solubilities in high temperatures brines. Geochimica et cosmochimica Acta, 49:195-203

Carroll, J.J. and Mather, A.E., 1989. The solubility of hydrogen sulphide in water from 0 to 90°C. Geochimica et Cosmochimica Acta, 53: 1163-1170

Drummond, S.E., 1981. Boiling and mixing of hydrothermal fluids chemical effects on mineral precipitation. Ph.D. Thesis, Pennsylvania State University, University Park, Pa.

Duan, Zhenhao, Moller, Nancy & Weare, John H., 1996. Prediction of the solubility of H_2S in NaCl aqueous solution: anequation of state approach. Chemical Geology 130 (1996) 15-20

Fernandez-Prini,R.; Alvarez, J. L. & Harvey, A. H., 2003. Henry's constants and vapor–liquid distribution constants for gaseous solutes in H_2O and D_2O at high temperatures. J. Phys. Chem. Ref. Data, Vol. 32, No. 2, 2003.

Fernández-Prini, R., and Crovetto, R.,1989. Evaluation of Data on Solubility of Simple Apolar Gases in Light and Heavy Water at High Temperature. J. Phys. Chem. Ref. Data 18, 1231.

Harvey, A. H., 1996. Semiempirical correlation for Henry's constants over large temperature ranges AIChE J. 42, 1491, 1996

IAPWS (International Association for the Properties of Water and Steam), 1995. Release on Values of Temperature, Pressure and Density of Ordinary and Heavy Water Substances at Their Respective Critical Points.In *Physical Chemistry of Aqueous Systems: Meeting the Needs of Industry* (Proceedings, 12th International Conference on the Properties of Water and Steam), edited by H. J. White, Jr., J. V. Sengers, D. B. Neumann, and J. C. Bellows ~Begell House, New York, 1995b!, p. A101.

Lee, J.I. and Mather, A.E., 1977.Solubility of hydrogen sulfide in water. Ber. Bunsenges Phys. Chem., 81: 1021-23.

Perez, Renee J. & Heidemann, Robert A., 2006. Coupling an equation of state and Henry's law to model the phase equilibria of gases and brines examples in the N_2–H_2O–NaCl system. Journal of Geochemical Exploration 89 (2006) 331–334

Perez, Renee J. & Heidemann, Robert A., 2007. Thermodynamic model for the solubility of light hydrocarbons in water and in aqueous sodium chloride solutions to elevated pressures and temperatures. Geofluids (2007)

Plyasunov, A. V.; O'Connell, J. P.; Wood, R. H. and Shock, E. L., 2000. Infinite dilution partial molar properties of aqueous solutions of nonelectrolytes. II. Equations for the standard thermodynamic functions of hydration of volatile nonelectrolytes over wide ranges of conditions including subcritical temperatures. Geochimica et Cosmochimica Acta, Vol. 64, No. 16, pp. 2779–2795, 2000

Setschenov, A., 1889. Action de l'acide carbonique sur les solutions des sels a acides forts. Annales de Chimie et de Physique 25, 226.

Soave, G., 1972. Equilibrium constants from a modified Redlich–Kwong equation of state. Chemical Engineering Science 27 (6), 1197–1203.

Using quartz and plagioclase to gain insight into chemical and thermal evolution of the Rotoiti magma prior to the caldera-forming eruption ±55 ka, New Zealand

Victoria Smith[1]*, Phil Shane[2], & Ian Nairn[3]

1. Department of Earth Sciences, University of Bristol, Wills Memorial Building, Queen's Road, Bristol BS8 1RJ, UK. * Victoria.Smith@bristol.ac.uk
2. School of Geography, Geology & Environmental Science, University of Auckland, Private Bag 92019, Auckland, New Zealand
3. 45 Summit Road, Rotorua RD5, New Zealand & GNS Science, Wairakei Research Centre, Taupo, New Zealand

Keywords: rhyolite, magma storage, magma mixing, quartz, plagioclase.

1. INTRODUCTION

The size of large silicic magma bodies and the generation of their compositional diversity seems to be inherently related to accumulation and the amount of mixing of the melts that enter the upper crust. Recent models suggest that magma bodies in the upper crust are incrementally assembled, with influx of new mafic magma melting relict crystalline bodies (e.g., Hildreth and Wilson, 2007). Evidence from plutonic bodies also suggests piecemeal assembly (e.g., Wiebe *et al.*, 2007).

Here we use the temporal record preserved in the crystals from the most recent caldera-forming eruption from the Taupo Volcanic Zone, New Zealand (Rotoiti) to investigate the processes that occurred prior to eruption, and provide insight into generation and assembly of silicic magmas within the upper crust. Plagioclase and quartz are ubiquitous in the deposits, and grow over a wide range of temperatures and pressures, thus were used to investigate the magmatic history Using micro-analytical techniques, profiles across both phases were carried out on crystals from all the deposits (see Shane *et al.*, In press and Smith *et al.*, Submitted for analytical methods). Ti compositional profiles across quartz used with the TitaniQ (see Wark *et al.*, 2006) geothermometer provide insight into the thermal history of the melts; while compositional variations across plagioclase provide a record of melt evolution.

2. GEOLOGICAL OVERVIEW

The Taupo Volcanic Zone (TVZ) in the central North Island of New Zealand is the actively rifting (~7 mm/yr) on-land continuation of the Tonga-Kermadec arc. Silicic eruption recurrence intervals are some of the highest on Earth, with >50 eruptions in the last 50 kyrs.

The Rotoiti eruption occurred ~55 ka from Okataina volcanic centre within the TVZ. More than 120 km^3 of magma was erupted, and led to collapse and caldera formation. Deposits have been divided into 3 stratigraphic units (Re1, Re2 and Re3). Most (>90%) of the rhyolite magma (T1) erupted during the Rotoiti eruption was low-K_2O, cummingtonite-bearing, with Fe-Ti oxide temperature of 766°C and oxygen fugacity of +0.92 with respect to nickel-nickel oxide buffer (NNO; Shane *et al.*, 2005). Deposits erupted post caldera collapse also have some pumices that are distinct, characterized by high-K glass, presence of biotite. This magma (T2) was also cooler and less oxidised (712°C; -0.16 NNO) than T1. Some mingled (T1+T2) clasts are found in the late deposits, which contain bimodal (disequilibrium) populations of all crystal phases, suggesting short-lived contact between the magmas (Shane *et al.*, 2005).

3. QUARTZ

Cathodoluminscence images of quartz from the deposits display distinct zoning (Fig. 1A). This zoning appears in most of the crystals, and is found in both magmas and throughout the eruption sequence. Abrupt increases in temperature/Ti concentration are common in many of the crystals (Fig 1B). The increases typically occur over distance <30 μm and correspond to temperature variations ~40°C. Crystals from throughout the sequence record these thermal events, suggesting the events were large. Quartz erupted in the early sequence (Re1) displays the least variation in Ti concentration/temperature (Fig. 2), while quartz found throughout the rest of the sequence (Re2 and Re3, both T1 and T2 magmas) displays significant variations, both within crystals and between crystals. Some crystals display ranges that do not overlap, indicating that some crystals crystallized at cooler temperatures than others. Quartz temperatures displaying ranges that extend to much lower temperatures than Fe-Ti oxide eruption temperatures (Fig. 2), suggesting a substantial rejeventation event postdated most quartz crystallization.

(A)

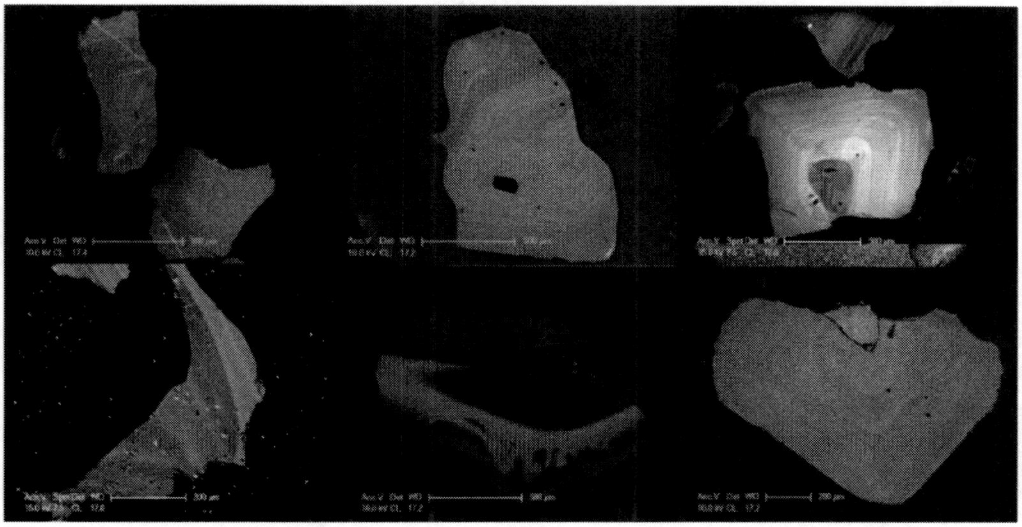

(B)

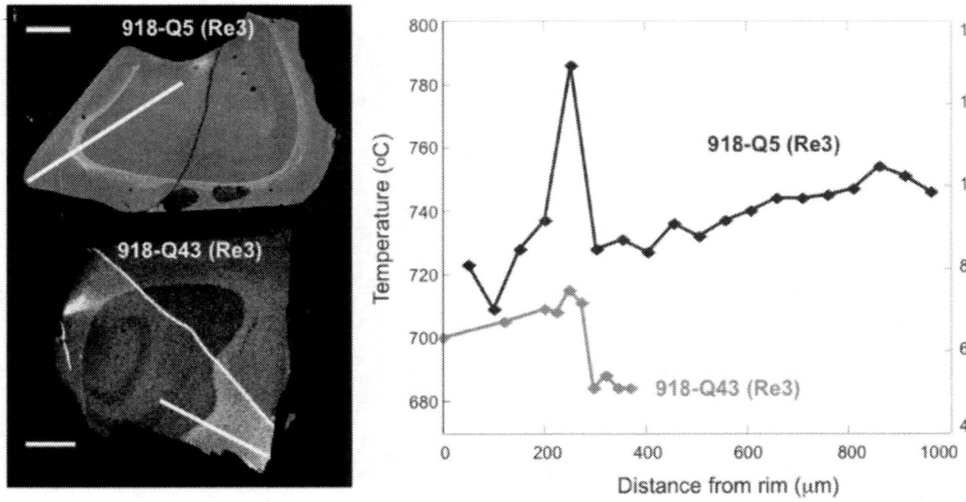

Figure 1: (A) Cathodoluminescene images of quartz crystals from the Rotoiti pyroclastics. Images on the left (top and bottom) correspond to crystals from Re1 (sample 910, T1 magma); middle to Re2 (sample 915; T1 magma), and on the right to Re 3 (sample 918, T2 magma). (B) Variations in Ti and temperature across quartz crystals in the T2 magma. Note

how greyscale directly reflects variation in Ti concentration. Some crystals display rapid changes in temperatures, and others display different temperature ranges.

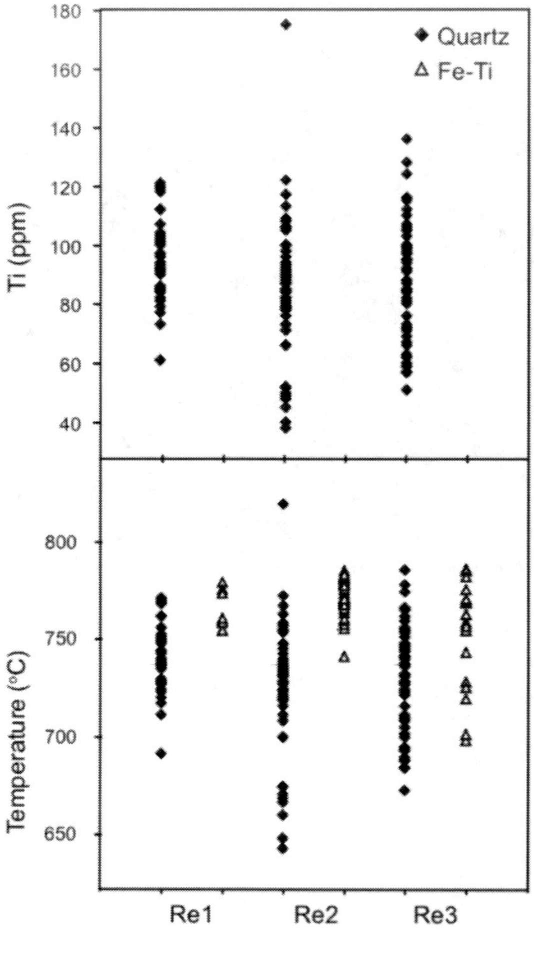

Figure 2. Range of temperatures from quartz versus Fe-Ti temperatures. Crystallisation of the quartz occurs over significantly wider range. The Fe-Ti temperatures represent temperature at time of eruption, as they are reset within weeks. Whereas, diffusion within the quartz crystals occurs over much longer timescales (>100 years), thus quartz provides a thermal record. It appears that crystallisation of the quartz generally occurred at temperatures that were lower than eruption temperatures, and indicates the magmas experienced thermal rejuvenatation prior to eruption.

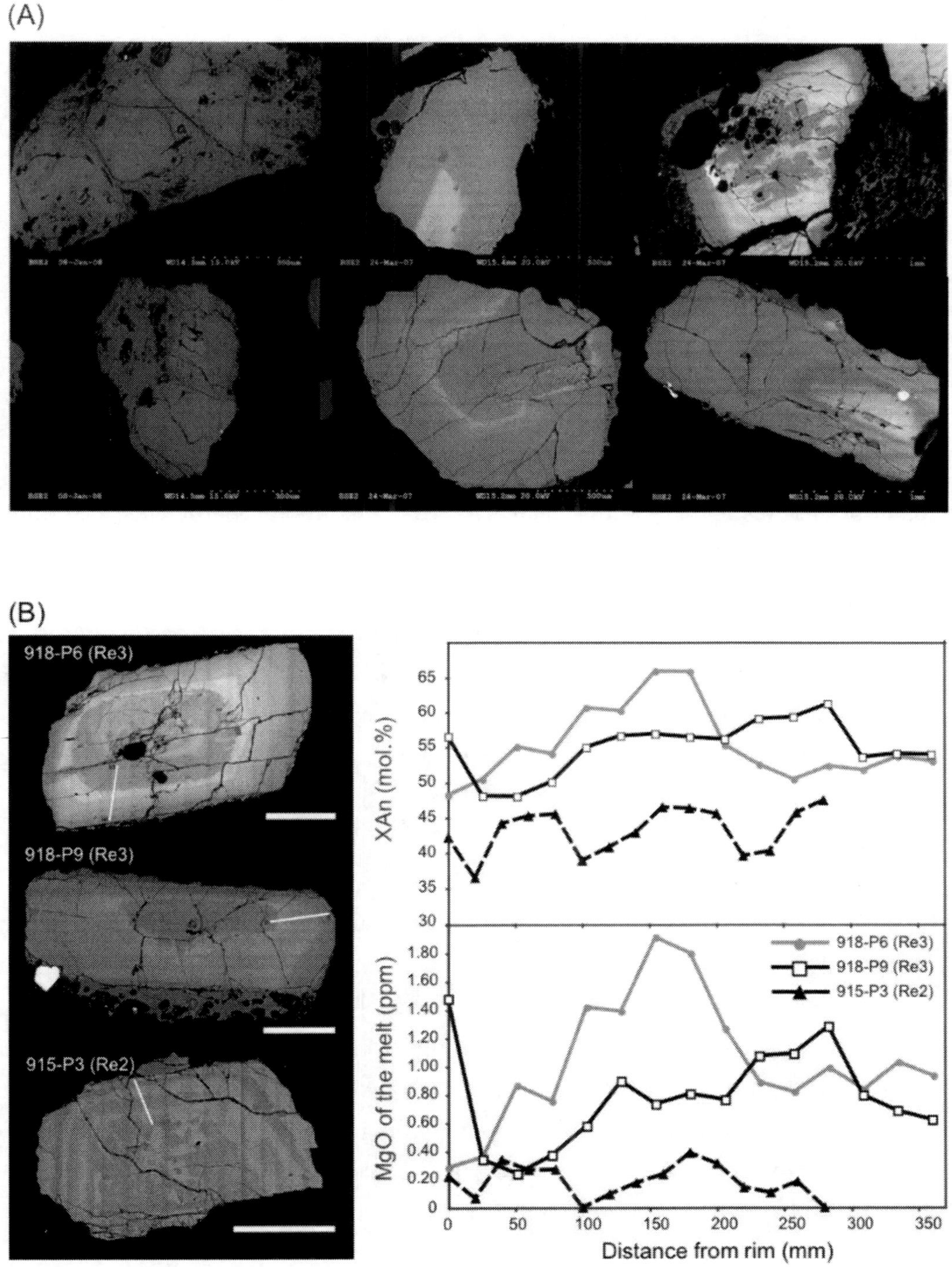

Figure 3: (A) Backscattered electron (BSE) images of plagioclase crystals in the Rotoiti pyroclastics. Images on the left (top and bottom) correspond to crystals from Re1 (sample 910); middle to Re2 (sample 915), and on the right to Re 3 (sample 918). (B) Profiles

IOP Conf. Series: Earth and Environmental Science **3** (2008) 012016 doi:10.1088/1755-1307/3/1/012016

across plagioclase crystals in the Rotoiti. High XAn and MgO_{melt} concentrations are observed in some crystals. These correspond to zones directly following a resorption horizon, they are direct evidence for recharge of the system.

4. PLAGIOCLASE

Plagioclase crystals show variations in both major (An_{36}-An_{68}) and trace elements (0-159 ppm MgO, 46-112 ppm TiO, 618-756 ppm Sr & 618-756 ppm Ba; Fig. 3). Variations observed in X_{An} (which BSE images reflect) are very sensitive to a range of parameters, e.g., temperature, pressure, H_2O and melt composition, which change considerably over time making it is difficult to dechipher magmatic processes. Instead more insight into magmatic processes can be gained from the record of melt evolution that can be reconstructed from trace element compositions (e.g., Smith *et al.*, Submitted). MgO melt compositions (MgO_{melt}) were calculated from partition coefficients determined by Bindeman *et al.* (1998). Calculated melt compositions display a wide range in MgO_{melt} (0-2 wt%; Fig 2B), much higher than the matrix glass composition (±0.12 wt%). The high MgO contents are recorded in specific zones that follow resorption horizons, and correspond to elevated X_{An} compositions. These notable increases in composition are direct evidence for recharge of the system by significantly more mafic melt. Plagioclase crystals, from samples analysed from Re2 and Re3 (encompassing both T1 and T2 magmas), display both textural and compositional evidence for a recharge event.

5. SUMMARY

The magmas that erupted during the Rotoiti event had a similar, complex thermal and compositional evolution. It appears that recharge events of hotter and more mafic magma commonly enter these upper crustal silicic systems. Recharge events are recorded as abrupt changes in crystal composition, suggesting limited diffusion, and indicating that crystals did not reside in melts in the upper crust for long periods of time (<100 years). These recharge events appear to be a mechanism of adding more melt to the system and allowing them to accumulate to such large volumes. They also appear to revive some stagnated magmas that have almost entirely cooled, and often trigger eruption.

REFERENCES

Bindeman, I.N., Davis, A.M. and Drake, M.J., 1998, Ion microprobe study of plagioclase-basalt partition experiments at natural concentration levels of trace elements: Geochimica et Cosmochimica Acta, 62, 1175-1193.

Hildreth, W. and Wilson, C.J.N., 2007, Compositional zoning of the Bishop Tuff: Journal of Petrology, 48, 951–999.

Shane, P., Nairn, I.A. and Smith, V.C., 2005, Magma mingling in the ~50 ka Rotoiti eruption from Okataina Volcanic Centre: implications for geochemical diversity and chronology of large volume rhyolites: Journal of Volcanology and Geothermal Research, 139, 295-313.

Smith, V.C., Blundy, J. and Arce, J.L., Submitted, A temporal record of magma accumulation and evolution beneath Nevado de Toluca, Mexico, preserved in plagioclase phenocrysts: Journal of Petrology.

Wiebe, R.A., Wark, D.A. and Hawkin, D.P., 2007, Insights from quartz cathodoluminescence zoning into crystallization of the Vinalhaven granite, coastal Maine: Contributions to Mineralogy and Petrology, 154, 439-453.

IOP Conf. Series: Earth and Environmental Science 3 (2008) 012017 doi:10.1088/1755-1307/3/1/012017

Ground deformation at collapse calderas: Influence of host rock lithology and reservoir multiplicity

A. Geyer and J. Gottsmann

Department of Earth Sciences, University of Bristol, Wills Memorial Building, Queen's Road, BS8 1RJ, Bristol, UK A.GeyerTraver@bristol.ac.uk

Keywords: caldera, ground, deformation, lithology, multiplicity

1. Introduction and objectives

One of the most common signals of caldera unrest is ground deformation. Various source mechanisms have been proposed to explain the observed deformation including magma intrusion, slip along faults, or pore pressure variations in transient hydrothermal systems (Dzurisin et al,. 2003; Poland et al., 2006; Masterlark, 2007). Since it is difficult to directly identify causative processes at depth, deformation sources need to be inferred by the analysis of observables. Numerical models provide one link between measured ground deformation and the inaccessible deformation source.

Here, we present a set of new results from numerical forward modelling. In our models we simulate the theoretically resultant vertical and horizontal surface displacements as the response of an essentially elastic medium to pressure changes in one or more shallow reservoirs with different geometries surrounded by host rock with variable mechanical properties. We show that medium heterogeneities such as layering of mechanically strong and weak rocks combined with the relative position of reservoirs may induce significant variations in the stress field that contribute to modify the ground deformation pattern at the free surface. Our findings are particularly relevant for flat-topped collapse caldera systems, where mechanical properties of caldera-fill successions are pronouncedly different to surrounding rocks and topography plays a minor role on stress distribution at the free surface.

2. Procedure

For each model the theoretical displacement and stress field has been obtained using FEMFES, a code that solves elasticity using a Finite Element Method with nodal implementation (Codina and Folch, 2004). The computational domain corresponds to an idealised cross-section of the upper crust of 50 km length, which extends to a depth of 25 km below a flat surface (Figure 1). Reservoirs are modelled as cavities and are characterized by two axes a and b, oriented in a Cartesian coordinate system x-y (y = depth). The upper planar surface of each of the models corresponds to the Earth's surface and is treated as a free surface (i.e. traction free). The inflating reservoir is represented by a uniform pressure acting outwards normal to the reservoir's surface. The computational margins are placed at vertical and radial distances times greater than the dimensions of the reservoir to avoid border effects.

On the one hand, we have performed numerical runs for one reservoir focused on studying the influence of variations in the mechanical properties of the host rock. We introduce mechanically different layers considering materials of high (stiff material) or low (soft materials) Young's modulus (E). Depending on the configurations of the different models (Figure 1), mechanical properties of these layers may corresponding to soft (E = 18 GPa, Poisson's ratio (ν) = 0.25) or stiff (E = 72 GPa, ν = 0.25) materials. We explore the effect on ground deformation when varying the thickness of the soft or the stiff layer, the ratio between the thicknesses of both layers, and the distance between them.

For all two-reservoir models, we are especially interested in the influence of distance between both reservoirs.

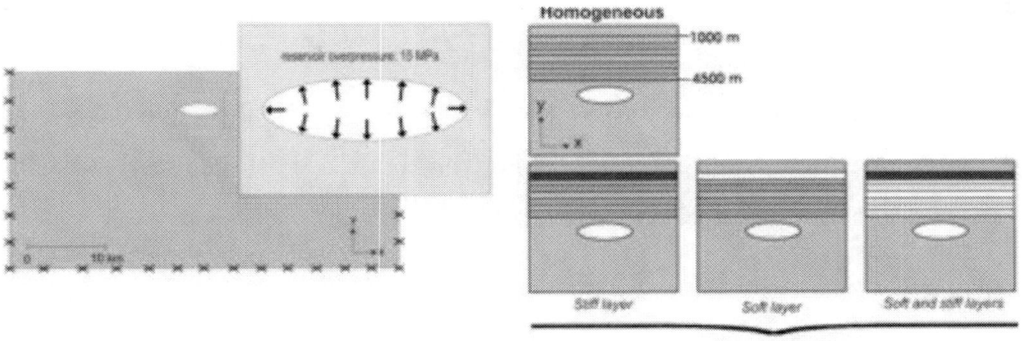

Figure 1: Representation of the computational domain (grey) with the reservoir located at a certain depth. Boundary conditions have been also illustrated. The reservoir has an

overpressure ΔP imposed at the reservoir walls. The Earth's surface is treated as a free surface (i.e. traction free) and the computational margins are fixed, i.e. zero displacement. In order to introduce heterogeneities in the model we define seven 500m thick layers between 1000m and 4500m depth. If the host rock is considered homogeneous, all layers have the same mechanical properties. Contrarily, by assigning different mechanical properties to the individual layers the host rock becomes heterogeneous.

3. Results

For those models exploring the influence of host rock heterogeneity, we compare theoretical vertical and horizontal ground displacement, u_{yy} and u_{xx}, with those obtained by the reference "HOMOGENOUS" model.

In general terms, compared to the "HOMOGENEOUS" result, the presence of mechanically different layers does neither modify the shape nor the footprint of the overall ground deformation. What is however changing is the amplitude of deformation depending on the thickness of the soft/stiff layers. Thick soft layers lead to a measurable increase in the amplitude, i.e. an increase in both the u_{yy}max and u_{xx}max values (Figure 2a). Contrarily, the presence of thick stiff layers reduces the amplitude, i.e. decreases u_{yy}max .

Concerning the influence of distance between soft and stiff layers, if the soft layers overlie the stiff layer, the amplitudes of both u_{yy} and u_{xx} are independent of the distance between the layers. In contrast, if stiff layers overlie soft layers, increasing the distance between both layers tends to increase u_{yy}max and u_{xx}max compared to the "HOMOGENOUS" model.

Concerning the models with two reservoirs, results indicate that, both the u_{yy} and the u_{xx} curves are strongly dependent of the horizontal distance between reservoirs Hdis. If Hdis increases, we find two separate u_{yy} peaks (Figure 2b).

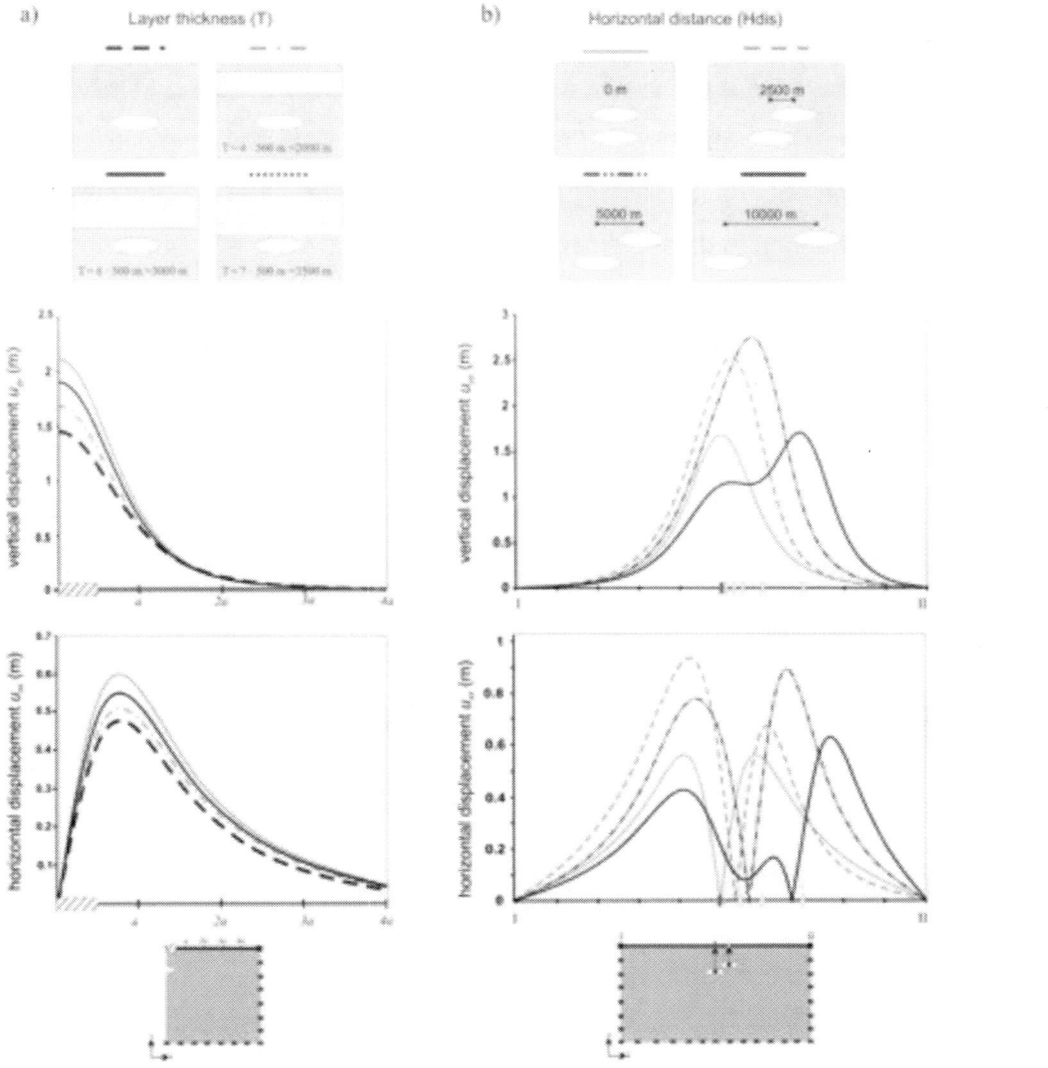

Figure 2: Theoretical vertical displacement u_{yy} horizontal displacement u_{xx} at surface. (a) Results for those numerical runs assuming different thicknesses for the soft layer **T**. (b) Results for those numerical runs assuming two reservoirs located at different horizontal distance (Hdis). The black and white rectangle indicates the location of the projection at surface of the deep and shallow reservoir centres, respectively.

4. Conclusions

The finite mechanical properties of the surrounding host rock and reservoir geometries measurably affect the resulting pattern of surface displacements and stress distributions around pressurised sources. The main points to highlight are: (i) the strong influence of soft and stiff layers and their mutual distribution on the amplitude of the surface deformation (both total vertical and horizontal displacements), (ii) the potential pitfall in assuming the ratio between vertical and horizontal displacements to be sensitive to reservoir geometries to explain ground deformation patterns in volcanic areas, (iii) the effect of reservoir multiplicity on the footprint of the deformation zone. The latter is an important observation, because a common assumption in most models of surface deformation to date, implicitly assume that the footprint of the deformed area is directly related to the source depth (Mogi, 1958). As shown here, this assumption is not of paramount validity as geometrically diverse pressure sources as well as mechanically heterogeneous host rocks can significantly alter the footprint of surface deformation above (a) pressurised source(s). As a consequence, stacking of causative sources (magmatic and hydrothermal) and mechanically distinct caldera-fill successions may result in multiple shareholders of resultant ground deformation during reactivation. Our study clearly identifies the need to better quantify mechanical properties of volcanic lithologies and detailed field analysis of caldera fill successions. The applicability of numerical models to shed light on subsurface dynamics during volcano deformation appears currently mostly confined by our limited knowledge of realistic mechanical properties of subsurface rocks and their distribution.

References

Codina, R., and Folch., A. (2004). "A stabilized finite element predictor-corrector scheme for the incompressible Navier-Stokes equations using a nodal-based implementation". International Journal of Numerical Methods for Fluids, 44: 483-503

Dzurisin, D., and D. A. Johnston. (2003). A comprehensive approach to monitoring volcano deformation as a window on the eruption cycle. Reviews in Geophysics 41: doi: 10.1029/2001RG000107.

Masterlark, T., 2007. Magma intrusion and deformation predictions: Sensitivities to the Mogi assumptions. Journal of Geophysical Research, 112(B06419): doi: 10.1029/2006JB004860.

Mogi, K., 1958. Relations of the eruptions of various volcanoes and the deformations of the ground surface around them. Bulletin of the Earthquacke Reseach Institute, 36: 99-134.

Poland, M., M. Hamburger, and A. Newman (eds). (2006). "The changing shapes of active volcanoes: History, evolution, and future challenges for volcano geodesy." Journal of Volcanology and Geothermal Research, The Changing Shapes of Active Volcanoes - Recent Results and Advances in Volcano Geodesy 150: 1-13. 1.

Influences of Magma Chamber Ellipticity on Ring Fracturing and Eruption at Collapse Calderas

Eoghan P. Holohan[1]*, Benjamin van Wyk de Vries[2], Valentin R. Troll[3], Thomas R. Walter[4], & John J. Walsh[1]

[1] Fault Analysis Group, School of Geological Sciences, University College Dublin, Belfield, Dublin 4, Ireland. Email: Eoghan.Holohan@ucd.ie

[2]Laboratoire Magmas et Volcans, 5 rue Kessler, 63038 Clermont-Ferrand, France.

[3] Department of Earth Sciences, Uppsala University, SE-752 36, Uppsala, Sweden.

[4] GFZ Potsdam, Telegrafenberg, Potsdam, D-14473, Germany.

Keywords: caldera, elliptical, subsidence, resurgence, ring fractures, volcanic eruptions.

1. ELLIPTICAL CALDERAS

Calderas are often elliptical in plan view, the most direct explanation for which is a similarly-shaped pre-caldera magma chamber. Indeed, plutons (relict pre-caldera magma chambers) are also often highly elongate in plan-view. Plan-view ellipticity of a pre-caldera magma reservoir, and its influence on the development of caldera ring fracturing and eruptive behaviour, have not previously been subjected to dedicated evaluation, however.

2. ANALOGUE MODELS OF ELLIPTICAL COLLAPSE AND RESURGENCE

In a first suite of experiments, we simulated caldera collapse into elliptical magma chambers (see Holohan et al. (2008) for details). A sand/gypsum mixture represented a brittle upper crust, and creamed honey replicated granitic magma (**Fig. 1**). The ratio of its horizontal long and short axes, A/B, defined the magma chamber's plan-view ellipticity, which we varied from 1.0 - 2.0. Draining the honey chamber through a basal conduit caused the chamber roof to collapse. A second, more limited suite of experiments simulated resurgence of collapsed elliptical magma chambers, into which we slowly injected honey back into the deflated chambers via the basal conduit.

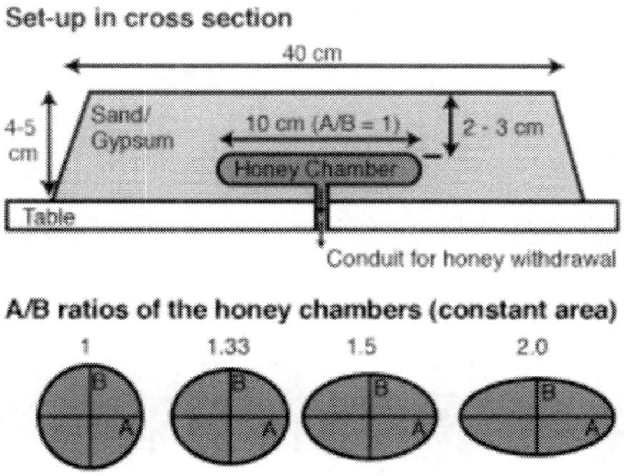

Figure 1: Sketch cross-section of experimental set-up.

3. MODEL RESULTS

Caldera collapse into highly-elliptical magma chambers (A/B = 2) produced a characteristic pattern of ring-fault localization and lateral propagation (Holohan et al. 2008). Two outward-dipping reverse faults localized, one around each end of the elliptical chamber's short axis (**Fig. 2a**), and then propagated to the ends of the roof's long axis (**Fig. 2 b-c**). The evolution of later inward-dipping normal ring-faults typically mirrored that of the reverse faults. Cross-sections reveal that central vertical displacement is accommodated more by faulting along the short axis and more by bending along the long axis (**Fig 2d-e**).

Although results are preliminary, the general deformation pattern for elliptical resurgence shows strong similarities to elliptical collapse. Ring faults accommodating uplift again initiate around the chamber's short axis (**Fig. 3a-c**) and are reverse, but dip inward (**Fig 3d-e**). They may be newly-generated reverse faults, inverted collapse-related normal faults, or a mix of both (**Fig 3d-e**). Again, central vertical displacement is accommodated more by faulting along the short axis and by bending along the long axis (**Fig 3d-e**).

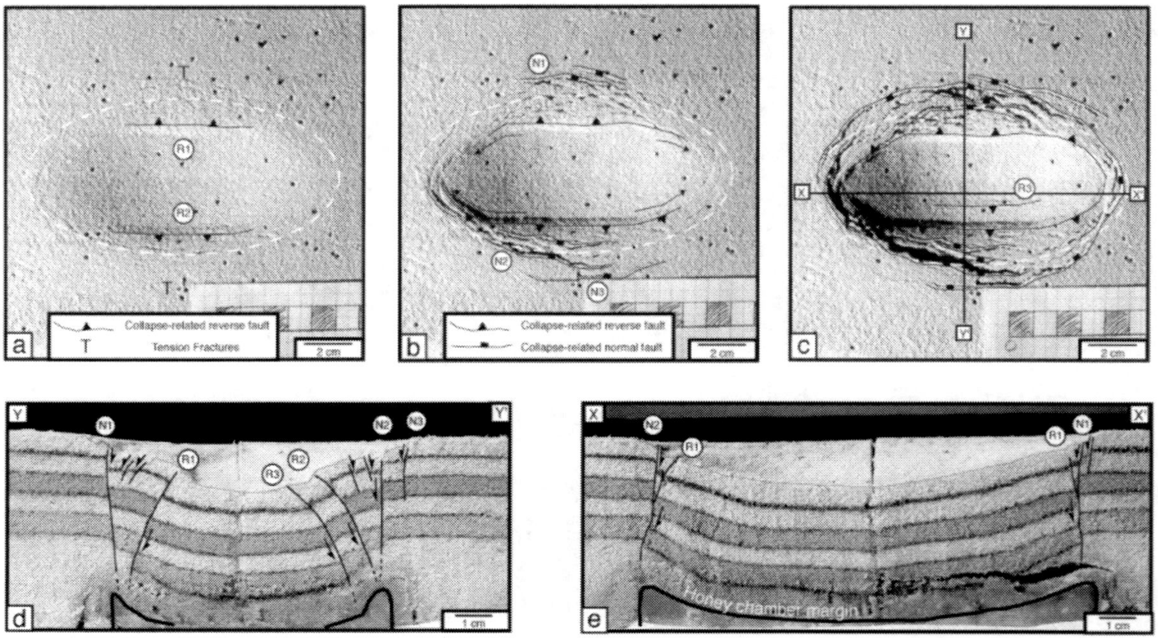

Figure 2: Results of highly elliptical caldera collapse models (A/B = 2.0). **(a-c)** Reverse and normal ring faults initiate at the short axis and propagate laterally to the long axis; **(d-e)** Cross-sections along the model's short and long axes respectively.

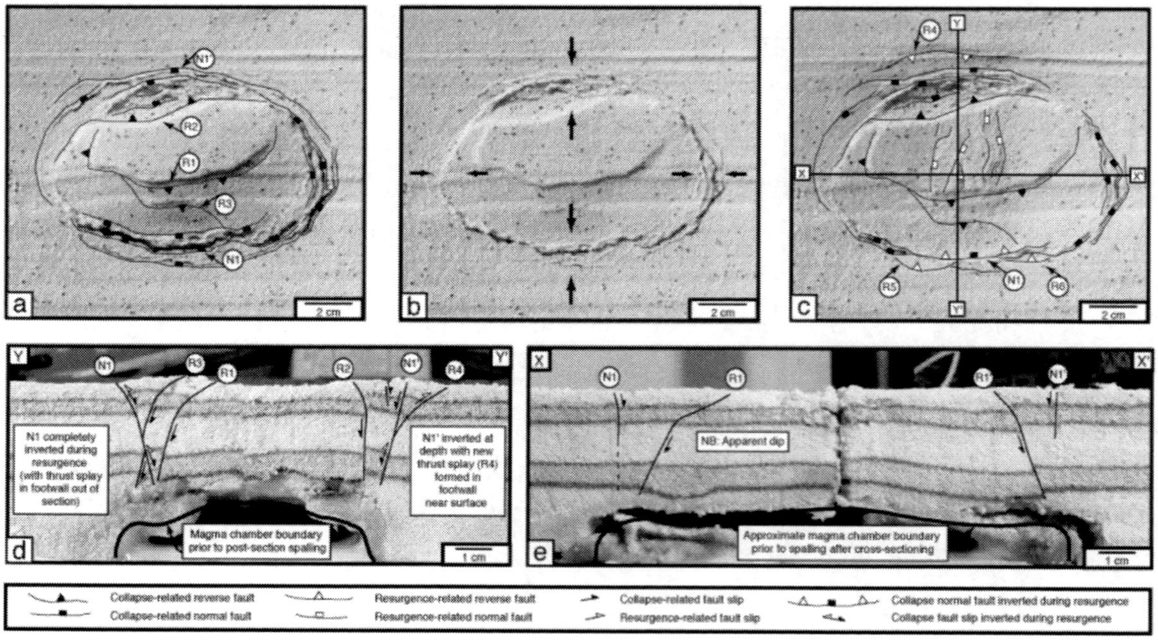

Figure 3: Results of highly elliptical caldera resurgence models (A/B = 2.0). **(a)** Elliptical caldera with main collapse-related faults highlighted; **(b)** Initial resurgence stage. Black arrows schematically indicate magnitude of horizontal shortening in marginal zones along long and short

axes; **(c)** Final resurgence stage, with resurgence-related faults highlighted; **(d-e)** Cross-sections along the model's short and long axes respectively.

4. RING FRACTURING AND ERUPTION AT ELLIPTICAL CALDERAS IN NATURE

Field and geophysical observations at several elliptical calderas of varying scale are consistent with a control from elliptical magma chamber geometry on ring fracturing and eruption, as predicted from our experiments. At Long Valley caldera **(Fig 4a)**, field studies (Wilson and Hildreth 1997) deciphered a pattern of syn-collapse ring-fracture localisation and lateral propagation ('unzipping') that closely matches that seen during highly-elliptical model collapse (cf. Holohan et al. 2008). At Katmai caldera **(Fig 4b)**, recent earthquake epicentres clearly cluster around the caldera's short axis and may define outward-dipping (and reverse?) ring-fault surface(s) (Moran 2003). At Rabaul caldera (**Fig 4c**), epicentre distributions approximate fault geometries seen in experiment: inward-dipping faults seem to branch at higher levels from a central outward-dipping ring fault. Vulcan vent lies along the caldera's short axis and on an inward-dipping fault, and erupted simultaneously with Tavurvur in 1994 after elliptical tumescence (Saunders 2001). This may reflect first inversion of collapse-related normal ring-faults around the short axis, in line with our model tumescence results (see **Fig. 3c-d**).

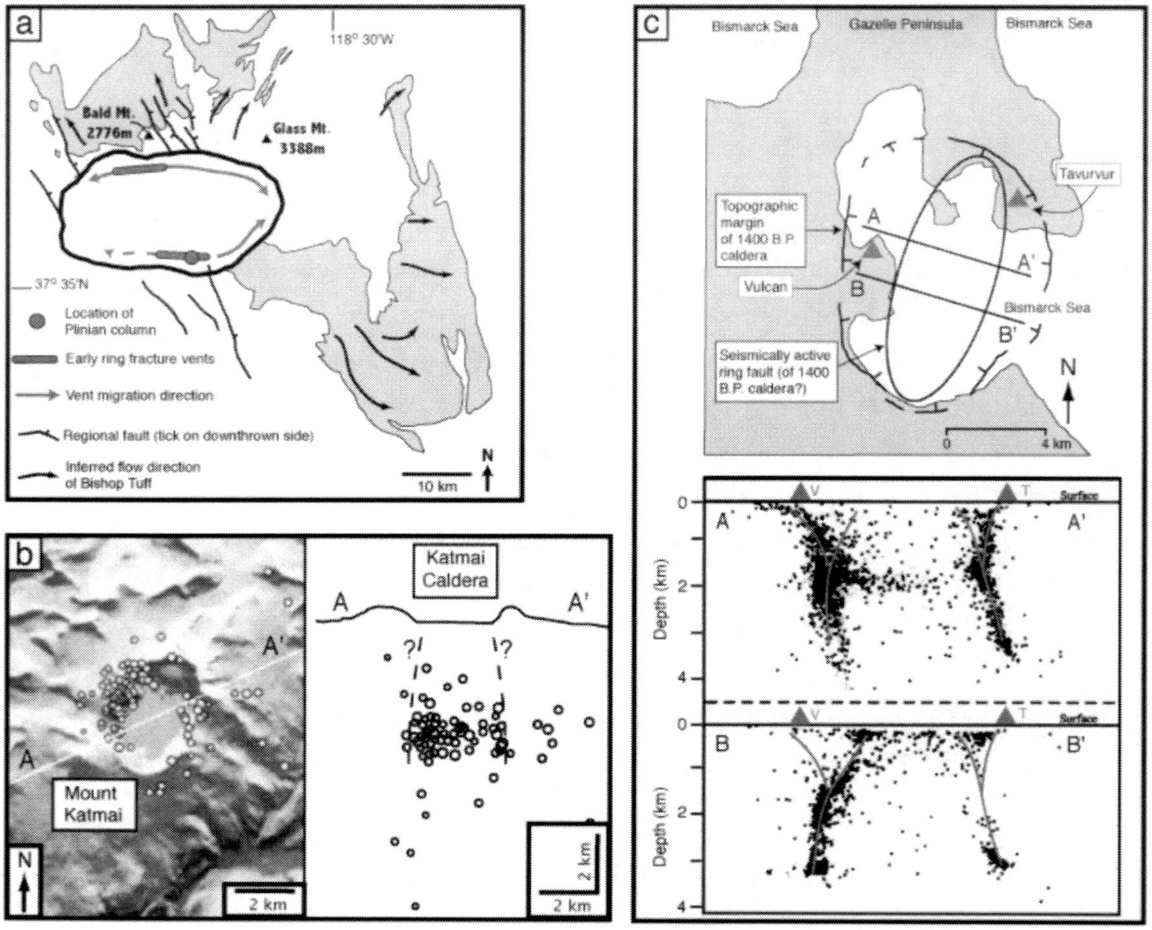

Figure 4: Evidence from natural calderas. **a)** Map of Long Valley caldera, California, and Bishop Tuff outcrop, with inferred pattern of initiation and migration of ring-fracture vents for successive ignimbrite packages (after Wilson and Hildreth 1997); **b)** Shaded relief map and cross-section of Katmai caldera, Alaska, with 'well-constrained' earthquake epicentres plotted or projected onto them (from Moran 2003); **c)** Sketch map of the 1400 BP Rabaul caldera, Papua New Guinea (after Nairn et al. 1995) with cross-section plots of earthquake epicentres (from Saunders 2001) and possible fault geometry as inferred from results of the present study.

5. DISCUSSION & CONCLUSIONS

Analogue models here demonstrate that magma chamber ellipticity may exert a strong control on patterns of ring fracturing and eruption at calderas, both during subsidence and tumescence/resurgence. Holohan et al. (2008) explained this as a result of bulk shear strain maximising along the chamber roof's short axis and leading to failure here first during subsidence. This principle should apply also with the inverse sense of deformation during elliptical tumescence/resurgence, a prediction supported here by experimental and field evidence.

REFERENCES

Holohan E.P., et al. (2008) Unzipping Long Valley: an explanation for vent migration during an elliptical ring fracture eruption: Geology, 36, 323-326.

Moran S.C., 2003, Multiple seismogenic processes for high-frequency earthquakes at Katmai National Park, Alaska: Evidence from stress tensor inversions of fault plane solutions: Bulletin of the Seismological Society of America, 93, 94-108.

Nairn, I.A. et al., 1995, Geology and eruptive history of Rabaul caldera area, Papua New Guinea: Journal of Volcanology and Geothermal Research, 69, 255-284.

Wilson, C.J.N., Hildreth, W.H., 1997, The Bishop Tuff; new insights from eruptive stratigraphy: Journal of Geology, 105, 407-439.

Saunders, S.J., 2001, The shallow plumbing system of Rabaul Caldera; a partially intruded ring fault?: Bulletin of Volcanology, 63, 406-420.

Effects of Host Rock Stratigraphy on the Formation of Ring-Faults and the Initiation of Collapse Calderas

Kinvig, H.S., Geyer, A., Gottsmann, J.

Department of Earth Sciences, University of Bristol, Wills Memorial Building, Queen's Road, BS8 1RJ, Bristol, UK

Keywords: caldera structure, ring fault, caldera, collapse

ABSTRACT

Most collapse calderas can be attributed to subsidence of the magma chamber roof along bounding sub-vertical normal faults (ring-faults) after a decompression of the magma chamber, following eruption. It has previously been shown that for ring-faults to initiate, and thus facilitate collapse, the stress field both at surface and around the magma chamber must satisfy specific critical conditions. Here, we present new numerical models that use a Finite Element Method to investigate the effects of variable stratigraphy (lithology/thickness/order of strata) on local stress field distribution. Results are compared with existing criteria for ring-fault initiation. Different subsurface scenarios were simulated by varying the stiffness (Young's modulus) of seven thin crustal layers placed above the magma chamber, and the host rock in which the chamber is seated. We consider the magma chamber to be subjected to a magmatic under-pressure of -15MPa, imposed at the chamber walls, in order to simulate magma withdrawal. Results indicate that for a given geometrical set-up, the magnitude and position of maximum tensional stress at the Earth's surface are influenced by the occurrence and relative distribution of mechanically stiff or soft lithologies above the magma chamber. For example, tensional stress at surface may be reduced by the presence of stiff layers (e.g. lavas), or increased by soft layers (e.g. pyroclastic units) compared to generic simulations using a homogeneous background medium. The observations suggest that the mechanical properties of crustal stratigraphy are therefore a further variable in the rare achievement of the stress field conditions required for ring-fault formation, and may be influential in generating or inhibiting caldera collapse. Therefore an understanding of pre-caldera stratigraphy could provide important insight into the likelihood of future caldera collapse events.

Generation of pyroclastic density currents from pyroclastic fountaining or transient explosions: insights from large scale experiments

Roberto Sulpizio, Pierfrancesco Dellino, Daniela Mele, Luigi La Volpe
CIRISIVU, c/o Dipartimento Geomineralogico, via Orabona 4, 70125, Bari, Italy;
r.sulpizio@geomin.uniba.it

Keywords: plume, column collapse, magma, experimental volcanology

1. INTRODUCTION

Pyroclastic density currents (PDCs) are among the most amazing, complex and dangerous volcanic phenomena. They are moving mixtures of particles and gas that flow across the ground, and originate in different ways and from various sources, during explosive eruptions or gravity-driven collapse of domes. They may be short-lived (highly unsteady) or relatively long-lived (sustained unsteady to quasi-steady) phenomena, driven by both magmatic or phreatomagmatic melt fragmentation (e.g., Cas and Wright, 1987; Carey, 1991; Branney and Kokelaar, 2002; Sulpizio and Dellino, 2008).

Pyroclastic density currents (PDCs) encompass a wide spectrum of eruption, transport and depositional processes and accordingly produce a variety of deposit types (e.g. Sparks, 1976; Druitt, 1998; Branney and Kokelaar, 2002; Sulpizio and Dellino, 2008).

Despite the abundance of studies concerning their behaviour, because of the complexity, the physics of PDCs are still poorly understood, making it difficult to define their transport and deposition processes. In particular, it is important to understand the generation mechanisms on the behaviour of PDCs, as this is a crucial parameter which influences the physics of gravity-driven flows (e.g. Branney and Kokelaar, 2002; Sulpizio et al., 2007; Sulpizio and Dellino, 2008).

There is increasing evidence, both from the stratigraphic analysis of past eruptions and from observation of recent explosive eruptions (Christiansen and Peterson, 1981; Rowley et al., 1981; Freundt and Schmincke, 1986; Chough and Sohn, 1990; Scott et al., 1996; Cole et al., 1998; Sulpizio et al., 2005; 2007), that PDCs are frequently generated during discrete impulsive events and not only by the continuous fountaining of a Plinian column. Therefore, in currents that initiate explosively, the behaviour of the PDCs critically relates to mechanisms of pyroclastic fountaining or of expansion of an overpressurised jet.

Both triggering mechanisms were investigated through an experimental facility for the generation of large-scale, multiphase, gravity-driven currents. The experiments here described are particularly devoted to understanding the inception and development of PDCs under such impulsive conditions by means of the fast application of a finite stress to a finite mass of pyroclastic particles via expansion of compressed gas (Dellino et al., 2007).

Figure 1: Sequential slides depicting the injection into the atmosphere and collapse of a pressure-adjusted pyroclastic jet.

Figure 2: Deposits of the two runs of experiments show in Figures 1 and 3. a) near impact deposits left by the main collapse in pressure-adjusted pyroclastic fountaining experiment. Note the hummocky, thick, and massive aspect of the deposits; b) overview of the same deposits as a); c) near impact deposits left by the collapses in overpressurised jet experiment. Note the thin deposits even very close to the impact area; d) overview of the same deposits as c).

2. RESULTS

Large scale experiments on PDC generation were performed using the facilities of Laboratorio Sperimentale di Vulcanologia at Dipartimento Geomineralogico, University of Bari (Dellino et al., 2007). Experimental runs were performed using fixed inflow pressure (175 bars) and pyroclastic load (200 kg), and changing the length of the conduit (from 55 to 320 cm), in order to investigate the collapsing behaviour of both pressure-adjusted and overpressurised pyroclastic jets. As an example, we presented here the results of two experimental runs performed using the 320 cm and the 55 cm-long conduit. The first run (conduit of 320 cm) can be considered as a proxy of a pressure-adjusted pyroclastic jet. Video-footage and image analyses, coupled with near field geophones and pressure sensors data allowed to shed light on behaviour of PDC-generating collapses. Figure 1 shows how the collapsing process can be divided in 3 main stages. The first stage starts with plug injection and encompasses the vertical motion of the pyroclastic mixture until the first sign of collapse (first row of slides in Fig. 1).

IOP Conf. Series: Earth and Environmental Science 3 (2008) 012020 doi:10.1088/1755-1307/3/1/012020

Figure 3: Sequential slides depicting the injection into the atmosphere and collapse of an overpressurised pyroclastic jet

At conduit exit the top part of the plug maintains a cylindrical shape due to the high particle concentration, while the lower part is more diluted and turbulent (Fig. 1). Gravitational forces and friction between the plug and the atmosphere govern the partition of kinetic energy transformation into potential energy and turbulence. In particular, turbulence partially disassemble the top and more concentrated part of the plug, but the almost constant section of the jet after injection testifies for the poor efficiency of this mechanism (Fig. 1). The formation of small collapses from the basal part of the jet indicates the onset of stage 2 (Fig. 1). Small collapses of pyroclastic material that rapidly transform into finger-jets characterise the whole stage 2. At the time that some of the small finger-jets start to hit the ground, the main part of the pyroclastic material entrapped in the top part of the jet starts to collapse, marking the onset of stage 3 (Fig. 1). The small finger-jets collapse in an area close to the conduit base, and originate small and diluted PDCs when reaching the ground. The main collapse generates a larger PDC, which rapidly expands radially phagocytizing the previous small PDCs (Fig. 1). The deposits are thick and massive close to the impact point and show a hummocky surface (Fig. 2a).

Figure 3 shows slides from the second run of experiments, which was performed using a conduit of 55 cm filled by pyroclastic material. This configuration is a proxy for a transient explosion that generates an overpressurised jet, since the ejecting material doesn't have the time for equalising with the atmospheric pressure. The experiment can be described in two stages. The first stage comprises the jet formation and expansion, while the second stage describes the collapse behaviour of the pyroclastic material. It is noteworthy that in this case the plug expands just after the ejection into the atmosphere, and rapidly triplicates the conduit radius (second slide in Fig. 3). The more diluted basal part of the jet escape laterally to the dense top part to form upward-directed finger jets (Fig. 3). In about 1 s the inner overpressure and turbulence completely disassembles the initial plug and expands up to ten time the conduit radius (Fig. 3). The formation of collapsing finger-jets marks the onset of stage 2. They develop at any level within the expanded cloud, and collapse almost simultaneously over a larger area than in the case of pressure-adjusted jet (Figs. 1 and 3). This collapsing behaviour produces a mass partition among many different, small-volume finger-jets, which results in generation of a number of small, diluted PDCs. The deposits are thin and widely distributed, even in the impact area (Figs. 2c and d), in agreement with the observed behaviour of PDC generation.

In summary, PDC generation from collapse of pressure-adjusted or overpressurised pyroclastic jets critically depends on behaviour of injection into the atmosphere, which controls the collapsing mechanisms and then the physical parameters of the initiating current.

Preliminary analyses of velocity and pressure data seem to confirm the quantitative data here presented, and will allow a complete and qualitative analyses of pyroclastic collapses in the near future.

REFERENCES

Branney, M. J., Kokelaar, P., 2002. Pyroclastic Density Currents and the Sedimentation of Ignimbrites. Geological Society, London, Memoirs, 27, 143 pp.

Carey, S. N., 1991. Transport and deposition of tephra by pyroclastic flows and surges. In: Sedimentation in Volcanic Settings, edited by R. V. Fisher and G. A. Smith. SEPM, Special Publications, 45, 39– 57.

Cas, R., Wright J. W., 1987. Volcanic Successions: Modern and Ancient. Allen and Unwin, St Leonards, NSW, Australia, 528 pp.

Chough, S. K., Sohn, Y. K., 1990. Depositional mechanics and sequences of base surges, Songaksan tuff ring, Cheju Island, Korea. Sedimentology, 37, 1115– 1135.

Christiansen, R. L., Peterson, D. W., 1981. Chronology of the 1980 eruptive activity. In: The 1980 Eruptions of Mount St. Helens, Washington, edited by P. W Lipman and D. R. Mullineaux. U.S. Geol. Surv. Prof. Pap., 1250, 17– 30.

Cole, P. D., Calder, E. S., Druitt, T. H., Hoblitt, R., Robertson, R., Sparks, R. S. J., Young, S. R., 1998. Pyroclastic flows generated by gravitational instability of the 1996– 1997 lava dome of Soufrière Hills Volcano, Montserrat. Geophys. Res. Lett., 25, 3425– 3428.

Dellino, P., Zimanowski, B., Büttner, R, La Volpe, L., Mele, D., Sul pizio, R., 2007. Large-scale experiments on the mechanics of pyroclastic flows: design, engineering and first results. J. Geophys. Res., 112, B04202, doi:10.1029/2006JB004313

Druitt, T.H., 1998. Pyroclastic density currents. In: Gilbert, J.S. and Sparks, R.S.J. (Eds), The Physics of Explosive Volcanic Eruptions. Geological Society, London, Special Publications, Vol. 145, pp.145–182.

Freundt, A., Schmincke, H. U., 1986. Emplacement of small-volume pyroclastic flows at Laacher See (East Eifel, Germany). Bull. Volcanol., 48, 39–59.

Rowley, P. D., Kuntz, M. A., Macleod, N. S., 1981. Pyroclastic-flow deposits. In: The 1980 eruptions of Mount St. Helens, Washington, edited by P. W Lipman and D. R. Mullineaux. U.S. Geol. Surv. Prof. Pap. 1250, 489– 512.

Scott, W. E., Hoblitt, R. P., Torres, R. C., Self, S., Martinez, M. L., Nillos, T. J., 1996. Pyroclastic flows of the June 15, 1991, climatic eruption of Mount Pinatubo. In: Fire and Mud: Eruptions of Mount Pinatubo, Philippines, edited by C. G. Newhall and S. Punongbayan, Philippine Institute of Volcanology and Seismology, Quezon City and University of Washington Press, Seattle, 545–570.

Sparks, R. S. J., 1976. Grain size variations in ignimbrites and implications for the transport at pyroclastic flows. Sedimentology, 23, 147– 188.

Sulpizio, R., Mele, D., Dellino, P., La Volpe, L., 2005. A complex, subplinian-type eruption from low-viscosity, phonolitic to tephri-phonolitic magma: The AD 472 (Pollena) eruption of Somma-Vesuvius (Italy). Bull. Volcanol., 67, 743– 767.

Sulpizio, R., Mele, D., Dellino, P., La Volpe, L., 2007. Deposits and physical properties of pyroclastic density currents during complex Subplinian eruptions: the AD 472 (Pollena) eruption of Somma-Vesuvius, Italy. Sedimentology, doi:10.1111/j.1365-3091.2006.00852.x.

Sulpizio, R., Dellino, P., 2008. Sedimentology, depositional mechanisms and pulsating behaviour of pyroclastic density currents. In: Gottsman, J., Marti, J., (Eds). Calderas volcanism: analysis, modelling and response. Developments in Volcanology, Vol. 10., 57-96. Elsevier, Amsterdam, doi 10.1016/S1871-644X(07)00002-2.

Types of collapse calderas

Aguirre-Díaz, Gerardo J.

Centro de Geociencias, Universidad Nacional Autónoma de México, Campus Juriquilla, Querétaro, Qro., 76230, Mexico. E-mail: ger@geociencias.unam.mx.

Keywords: caldera, structure, collapse, magma

1. Introduction

A collapse caldera is a volcanic structure that forms during the collapse of crustal blocks on top of a shallow magma chamber (Cas and Wright, 1987; Lipman, 1997; Martí et al., 1994; Cole et al., 2005; Gottsmann and Martí, 2008). During this collapse, a large volume of magma is evacuated explosively in the form of pyroclastic fallouts and pyroclastic flows (Druitt and Sparks, 1984). In general, lava domes and cones are emplaced after collapse along the caldera structural margin and/or inside the caldera (e.g., Bailey et al., 1976; Aguirre-Díaz, 1996). The result of this catastrophic explosive volcanic collapse is a depression that could end with different shapes, circular, oval, rectangular, or irregular, depending on the different factors involved in the caldera process. These factors can be internal, such as those related to the sub-caldera magma chamber (pressure changes, magma composition, magma inputs, magma rheology, etc.), or external, such as the stress or tectonic regime affecting the volcanic field or province where the caldera is developed.

2. Types of collapse calderas

Three main types of collapse calderas can be defined, 1) summit caldera, 2) classic caldera, and 3) graben caldera.

1) *Summit calderas.-* Summit calderas are those formed at the top of large volcanoes and are related to relatively small-volume pyroclastic products that include plinian fallouts and ignimbrites, such as Crater Lake, Ceboruco, Las Cañadas, and Somma-Vesuvio.
2) *Classic calderas.-* Classic calderas are semi-circular to irregular-shaped large structures, several km in diameter that are related to relatively large-volume pyroclastic products that started generally with pumice fallouts and continued

with the eruption of large pyroclastic flows that formed widespread ignimbrites. Examples of this type of caldera are Long-Valley, Yellowstone, Campi Flegrei, Valles, and Los Humeros.

3) *Graben calderas.*- Graben calderas are explosive volcano-tectonic collapse structures from which large-volume, ignimbrite-forming eruptions occurred through several fissural vents along the graben master faults and the intra-graben block faults, causing the collapse of the whole graben or of a sector of it. The main products of graben calderas are surge-deposits and large-volume widespread ignimbrite sheets. Pumice fallouts are practically absent. Examples include the Sierra Madre Occidental in Mexico, La Pacana (Andes), Toba (a pull-apart graben?), calderas of the Catalan Pyrenees, and perhaps Scafell (U. K.) and Taupo (New Zealand). These are probably the largest of the three types of calderas and the source of the most voluminous ignimbrites in the world.

3. Style of collapse

Following Lipman (1997), any of the three caldera types mentioned above could have collapsed at least in three different ways:

1) *Piston.*- When the collapse occurs as a single crustal block.
2) *Trap-door.*- When collapse occurs unevenly along one side while the opposite side remains with no collapse.
3) *Piece-meal.*- When collapse occurs as broken pieces of the crust on top of the magma chamber.

Therefore, depending on the caldera type and collapse style, different combinations can be applied to classify a caldera; for instance, a piston summit caldera, a trap-door classic caldera, or a piece-meal graben caldera.

IOP Conf. Series: Earth and Environmental Science **3** (2008) 012021 doi:10.1088/1755-1307/3/1/012021

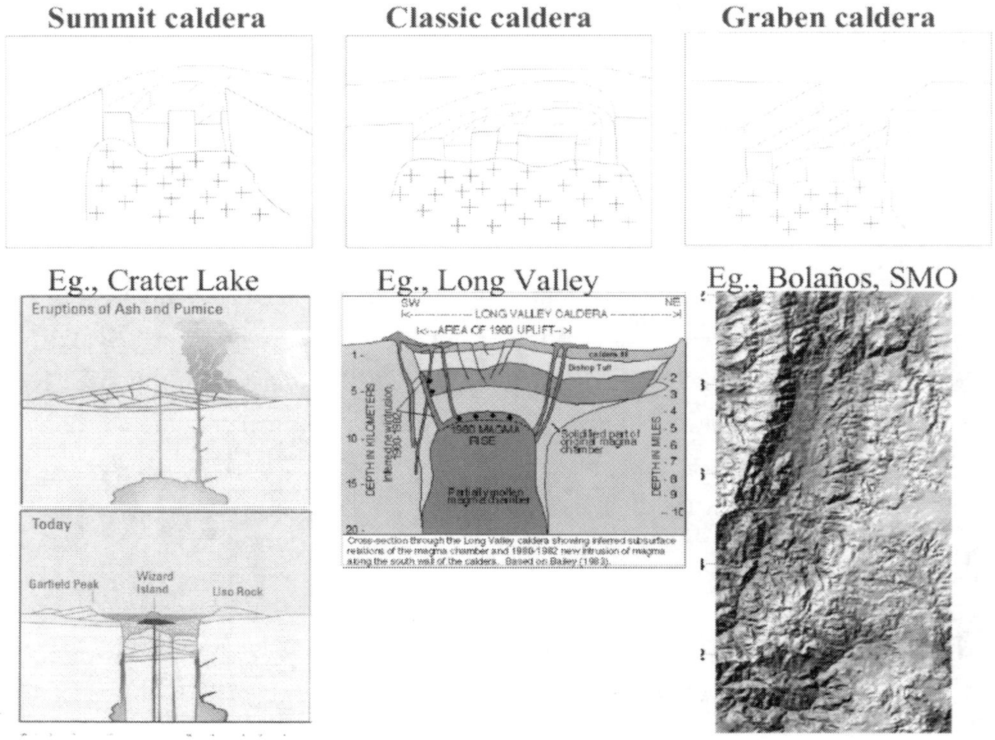

Figure 1. Schematic representation of the three main types of collapse calderas, summit, classic and graben, showing a corresponding caldera example.

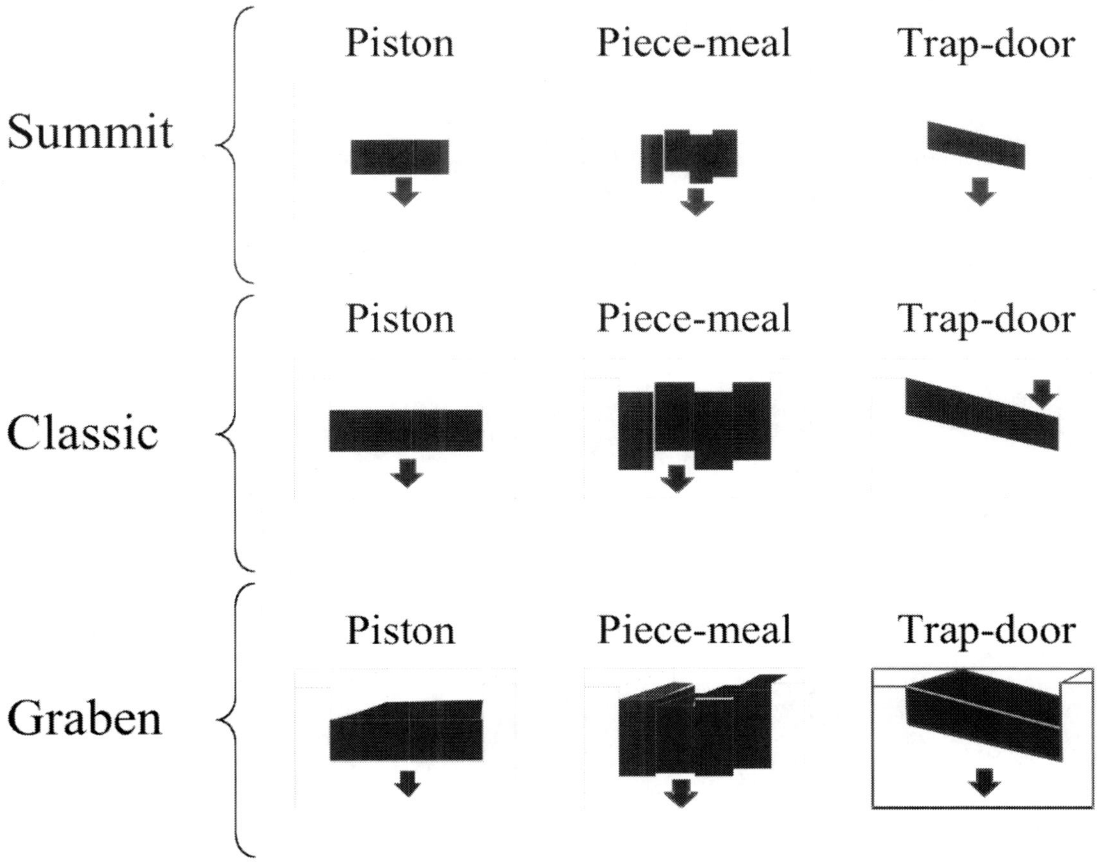

Figure 2. Classification of collapse calderas according to the different types and different styles of collapse.

References

Aguirre-Díaz, G. J., 1996, Volcanic stratigraphy of the Amealco caldera and vicinity, Central Mexican Volcanic Belt. Revista Mexicana de Ciencias Geológicas, v. 13, p. 10-51.

Bailey, R. A., Dalrymple, G. B., and Lanphere, M. A., 1976, Volcanism, structure, and geochronology of Long Valley Caldera, Mono County, California, Journal of Geophysical Research, v. 81 , p. 725-744.

Cas, R.A.F. and Wright, J.V., 1987, Volcanic Successions. Modern and ancient. Allen & Unwin, London, 528 p.

Cole, J.W., Milner, D.M., and Spinks, K.D., 2005, Calderas and caldera structures: a review. Earth Science Reviews, v. 69, p. 1–96.

Druitt, T.H. and Sparks, R.S.J., 1984, On the formation of calderas during ignimbrite eruptions. Nature, v. 310, p. 679–681.

Gottsmann, J. and Martí, J., 2008, Caldera Volcanism: Analysis, Modelling and Response, Developments in Volcanology 10, Elsevier, Amsterdam, 492 p.

Lipman, P.W., 1997, Subsidence of ash-flow calderas: Relation to caldera size and magma-chamber geometry: Bulletin Volcanologique, v. 59, p. 198–218.

Martí, J., Ablay, G.J., Redshaw, L.T., and Sparks, R.S.J., 1994, Experimental studies of collapse calderas. Journal Geol. Soc. Lond., v. 151, p. 919–929.

IOP Conf. Series: Earth and Environmental Science **3** (2008) 012022 doi:10.1088/1755-1307/3/1/012022

Piston to funnel - successive growth of a collapsed caldera during the Miyakejima 2000 eruption

Nobuo Geshi (Geological Survey of Japan, AIST, 1-1-1 Higashi, Tsukuba Ibaraki 305-8567 Japan, geshi-nob@aist.go.jp

Keywords: collapse, caldera, funnel-type caldera, magma chamber, Miyakejima

1. MIYAKEJIMA 2000AD caldera

The successive growth of the Miyakejima 2000 caldera exhibits the structural development of the collapse caldera from a piston-type in the early stage to a funnel type in the later stage as the progress of the subsidence. The Miyakejima 2000 caldera was formed by the collapse of a magma chamber at around 3 km depth by the evacuation of magma through the dike intruded beneath the NW off of the volcano (Kumagai et al., 2001; Geshi et al., 2002). The summit caldera appeared on July 8, 12 days after the beginning of the unrest and the caldera subsidence continued for 40 days contemporaneous with the cyclic seismic swarm with very-long period tremor (Kumagai et al., 2001). The cyclic subsidence of the caldera floor enlarged the volume of the caldera about 1.4×10^7 m^3/day for about 40 days. The size of caldera finally reached 1.7 km across, 500 m deep and 6×10^8 m^3. The total volume of the ejecta produced by the series of phreatomagmatic eruptions took place during and immediate after the caldera growth is less than 2 vol% of the collapse volume of the caldera.

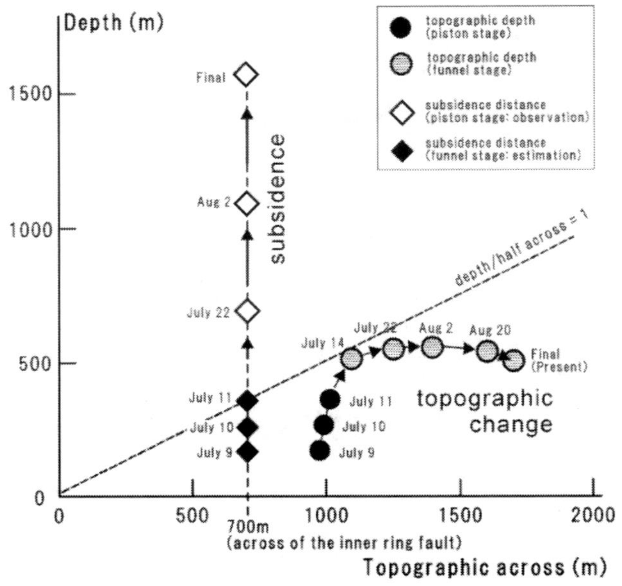

Figure 1: Structural development of the 2000 caldera. Solid circle: the topographic depth in the piston stage. Patterned circle; the topographic depth in the funnel stage. Solid diamond: the observed subsidence distance of the coherent piston. Open diamond: the estimated subsidence distance from the caldera volume and the area surrounded by the ring faults on the caldera floor.

2. Caldera development

The initial caldera has a concentric fault system. The coherent block surrounded by the ring faults subsided almost vertically within the first three days. This structure corresponds to the "piston" caldera of Lipman (1998). The concentric fault system consisted of the inner reverse faults surrounding a coherently-subsided block at the central portion of the caldera and the outer normal faults bordered the outline of the caldera. The subsided block at the center remained the original ground surface on the top of the block. Within the first three days from the caldera appearance, the depth of the caldera increased about 80 m/day with the subsidence of the central coherent block.

In the following stage, the subsiding block at the central portion of the caldera was completely covered with the deposits of the debris avalanches from the outward-migrating caldera wall. Though the total volume of the caldera increased almost constantly during this stage, the topographic depth of the caldera remained about 500 m from the original surface (Fig. 1) with the balance between the subsidence of the caldera floor and the filling by the landslides from the surrounding caldera wall (Geshi et al., 2002). The ephemeral circular faults with about 700 m across were observed on the surface of the debris during this period showing the continuous subsidence of the central portion of the caldera. The ratio between the caldera volume and the area of the subsided block surrounded by the ring fault indicates the total subsidence distance was 1.6 km or more (Fig. 1). This means that the thickness of the cover on the subsided piston exceeds 1 km. The final structure of the caldera has a small caldera floor surrounded by a steep caldera wall and this is akin to the funnel caldera of Lipman (1998).

3. Piston to funnel

The structural development observed in the case of Miyakejima indicates that the funnel-type caldera can develop from the piston-type caldera without explosive eruption. The successive subsidence enlarges the ratio of the structural depth (subsidence distance) against the horizontal across of the piston and this process enhances the instability of the caldera wall. The instability of the caldera wall will induce the landslides and collapse of the caldera wall and, consequently, the horizontal across of the caldera enlarged beyond the area surrounded by the ring fault. The outward migration of the caldera wall almost stopped imminently after the cessation of the cyclic seismic swarm indicating that the instability of the caldera wall induced by the subsidence of the caldera floor. The deposits derived from the caldera wall buried the subsiding piston at the center of the caldera and the structural cross-section of the caldera closes to the funnel-shape with a small caldera floor and a steep wall. In the case of Miyakejima, the transition from the piston-type to funnel-type occurred when the ratio of the subsidence against the half across of the ring fault beyond 1.

The shape of the magma chamber is also important factor to form a funnel-type caldera. The magma chamber should have large vertical size to allow the deep subsidence of a coherent block to form a funnel-type caldera. A spindle-shaped magma chamber with small horizontal across and large vertical extension is suitable for the funnel-type caldera because its small horizontal size will restrict the horizontal size of the subsiding piston and its large vertical size will allow the large vertical subsidence of the coherent piston into the magma chamber.

References

Geshi, N., Shimano, T., Chiba, T. Nakada, S., 2002, Caldera collapse during the 2000 eruption of Miyakejima Volcano, Japan. Bulletin of Volcanology, 64, 55–68. doi: 10.1007/ s00445-001-0184-z

Kumagai, H., Ohminato, T., Nakano, M., Ooi, M., Kubo, A., Inoue, H., Oikawa, J., 2001, Very-long-period seismic signals and caldera formation at Miyake Island, Japan. Science, 293, 687-690.

Lipman, P.W., 1997, Subsidence of ash-flow calderas: relation to caldera size and magma-chamber geometry. Bulletin of Volcanology, 59, 198-218.

Material toughness, internal structure, and caldera-collapse frequencies in basaltic and composite edifices

Agust Gudmundsson

Department of Earth Sciences, Queen's Building, Royal Holloway University of London, Egham TW20 0EX, UK (e-mail: a.gudmundsson@es.rhul.ac.uk)

Keywords: material toughness, fracture, structure, rock mechanics, caldera

Abstract

Why is caldera collapse, including slip on existing calderas, so much more common in basaltic edifices (shield volcanoes) than in composite edifices (stratovolcanoes and evolved calderas)? Since large landslides and dyke-fed eruptions are also apparently more common in basaltic edifices than in composite edifices, the question can be generalised as follows: Why is it easier to propagate a large-scale fracture in a basaltic edifice than in a composite edifice? To answer this question, I focus on three related topics: (1) fracture initiation and propagation, (2) material toughness, and (3) the internal structure of the edifices. An active volcano undergoing deformation generates and propagates numerous fractures. However, most of these soon become arrested and thus remain short. Short fracture cannot generate collapse calderas, large landslides, or dyke-fed eruptions; only large fractures can, namely fractures that propagate through many layers and reach the surface of the volcano. Most fractures become arrested and remain short because after a short propagation they meet with unfavourable mechanical conditions, particularly at contacts between layers. The mechanical conditions at contacts are primarily controlled by three parameters, namely: (a) local stresses; (b) tensile (or shear) strength of the contact versus that of adjacent layers; and (c) material toughness of the contact versus that of the adjacent rock layers.

Parameters a-c all contribute to the edifice material toughness, which is also referred to as critical strain energy release rate. It is a measure of resistance to fracture; it is the energy absorbed in a rock per unit area of fracture and has the units J m^{-2}. Thus, in a tough edifice, large amounts of energy are needed to cause failure through fracture propagation. Using field observations as well as conceptual and numerical models, I show that it is normally much more difficult for a large fracture such as a caldera fault, a ring dyke, or a feeder-dyke to propagate (through many layers) to the surface in a composite edifice than in a basaltic edifice. This difficulty is related to difference in internal structure between these edifice types. A composite edifice is a structure made of layers, rock units, and contacts whose mechanical properties vary widely. By contrast, a basaltic edifice is made of rock units, layers, and contacts whose properties vary much less. A large elastic mismatch (difference between Young's moduli) normally encourages fracture arrest or deflection at contacts, and thus increases edifice toughness. Also, unfavourable local stresses commonly develop at contacts between mechanically dissimilar layers. Thus, a composite edifice tends to arrest its fractures while they are short; it is therefore tougher and more resistant to any type of large-scale fracture propagation than a basaltic edifice. I suggest this difference in material toughness between these types of edifices is one principal reason for their difference in caldera (and landslide and feeder-dyke) failure frequencies.

Location of Silicic Caldera Formation in Arc Settings

Gwyneth R. Hughes and Gail A. Mahood
Department of Geological and Environmental Sciences, Stanford University, 450 Serra, Mall, Building 320, Stanford, CA 94305-2115, USA

Keywords: Calderas, silicic volcanism, arc volcanism

1. Introduction

Large-scale silicic magmatism is frequently observed in the geological record as plutons or ignimbrites, and yet the processes that produce silicic magma chambers are frequently under discussion. Silicic calderas represent the surface expressions of silicic magma chambers, and thus the study of their spatial distribution may give insight into the tectonic and crustal factors that favor the formation of evolved magmas. Of particular interest is the occurrence of silicic calderas within arc settings, because silicic magmatism at margins is commonly thought to stabilize continental crust. The goal of this study is to determine whether silicic calderas consistently occur "behind" the volcanic front, defined by the stratovolcanoes and shield volcanoes closest to the trench. This phenomenon has been observed, for example, in Mexico (Campos-Enriquez *et al.*, 2005), northern Central America (Rose *et al.*, 1999), and the Central Andes (Seibel *et al.*, 2001), but this study is the first to test whether it is universally the case.

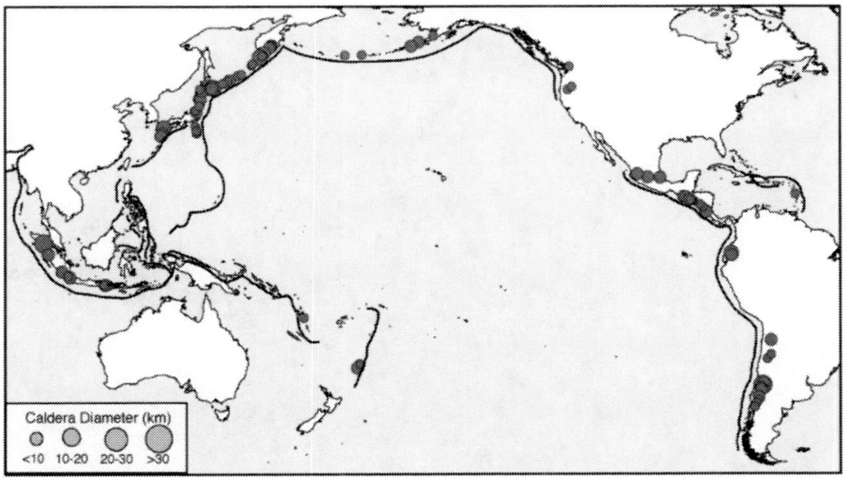

Figure 1: Map of the included silicic calderas in the 17 "simple" subduction zones.

2. Results

We examined 91 calderas in 17 arcs with simple tectonic settings (Figure 1). The calderas used in this study were taken from the compilation by Hughes and Mahood (in revision). All calderas were silicic (greater than 63 wt% SiO_2), larger than 5 km in diameter, and younger than 2 Ma. Arcs were categorized as oceanic (*e.g.* Aleutians), continental (*e.g.* Andes), or transitional-continental (*e.g.* Sumatra). For comparison, we also examined the locations of shield volcanoes and stratovolcanoes, as listed by the Smithsonian Global Volcanism Project (Siebert and Simkin, 2002). Each arc was subdivided into sections 100 km long, and the perpendicular distance from each caldera and volcano in the section was measured to the trench. The distance of the non-caldera volcano closest to the trench was then subtracted from both the caldera and volcano distances. The resulting distances (now with respect to the volcanic front) for caldera and non-caldera volcanoes were then compared.

The most striking result (Figure 2) is the spread of silicic calderas in continental margins over a wide range (from < 10 km to ~90 km) behind the volcanic front. In contrast, all the calderas in oceanic arcs and almost 80% of calderas in arcs with transitional continental crust are located within 20 km of the volcanic front. Occurrence of non-caldera volcanoes in all three arc types decays steadily and rapidly with distance from the front. We conclude then that silicic calderas in continental arcs do form preferentially behind the main volcanic axis.

IOP Conf. Series: Earth and Environmental Science **3** (2008) 012024 doi:10.1088/1755-1307/3/1/012024

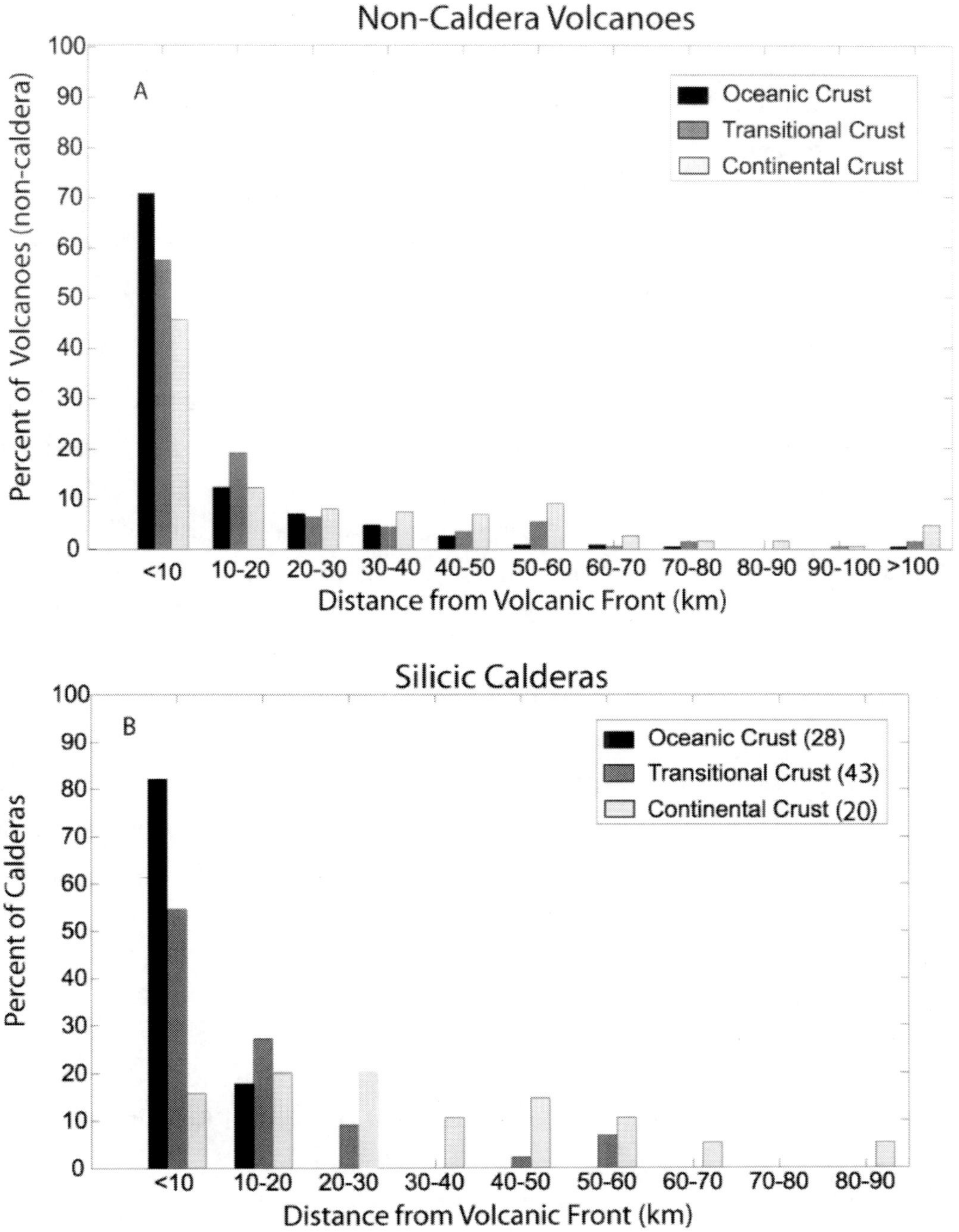

Figure 2: Histograms of the distances of intermediate and mafic volcanoes (non-caldera), and silicic calderas behind the volcanic front. Calderas have been divided into three groups based on the nature of the underlying crust: oceanic, transitional continental (*i.e.* underlain by a continental "chip". a) Stratovolcanoes and shield volcanoes; b) Silicic calderas.

3. Interpretation and Conclusions

There are several possible interpretations of this result. First, many studies, most recently Lallemand *et al.* (2005), find that slab dip is shallower beneath continental margins than oceanic margins. Volcanic arcs above shallower subduction zones would tend to be wider, and thus observed differences might simply be due to differences in arc width. The result, however, that silicic calderas in continental settings are located over a wide region behind the arc, whereas the number of non-caldera volcanoes in such settings decays rapidly away from the volcanic axis, suggests that the story is more complicated.

A second explanation could be that in oceanic arcs, crust that has evolved by repeated intrusion and partial melting will underlie only the main arc (Tamura and Tatsumi, 2002), where basaltic input is at a maximum. Several studies, (*e.g.* Hildreth and Moorebath, 1988) show that the composition of underlying crust has a profound influence on magmatic composition. Thus silicic calderas in oceanic arcs may only occur in the zone where oceanic crust has evolved by processes of continuous arc magmatism. Finally, during the course of its protracted lifetime, continental crust develops structures such as sutures, fault zones and graben that might channel magma flow and influence whether magma is stored in the upper crust rather than episodically erupted. This accumulated storage time is necessary for the development of large volumes of silicic magma. It is certainly possible that the locations of silicic calderas on continental crust are determined by some combination of these three factors.

References

Campos-Enriquez, J. O., Dominguez-Mendez, F., Lozada-Zumaeta, A., Morales-Rodriguez, H. F. and, Andaverde-Arredondo, J. A., 2005, Application of the Gauss theorem to the study of silicic calderas: The calderas of La Primavera, Los Azufres, and Los Humeros (Mexico): Journal of Volcanology and Geothermal Research, 147(1-2), 39-67.

Hildreth, W. and, Moorbath, S., 1988, Crustal contributions to arc magmatism in the Andes of Central Chile: Contributions to Mineralogy and Petrology, 98(4), 455–489.

Lallemand, S., Heuret, A. and, Boutelier, D., 2005, On the relationships between slab dip, back-arc stress, upper plate absolute motion, and crustal nature in subduction zones: Geochemistry, Geophysics, Geosystems, 6(9), 1-18.

Rose, W. I., Conway, F. M., Pullinger, C. R., Deino, A. and, MacIntosh, W. C., 1999, An improved age framework for late Quaternary silicic eruptions in northern Central America: Bulletin of Volcanology, 61(1-2), 106-120.

Siebel, W., Schnurr, W. B. W., Hahne, K., Kraemer, B., Trumbull, R. B., van den Bogaard, P. and, Emmermann, R., 2001, Geochemistry and isotope systematics of small- to medium-volume Neogene-Quaternary ignimbrites in the southern Central Andes: Evidence for derivation from andesitic magma sources: Chemical Geology, 171(3-4), 213-237.

Siebert, L. and, Simkin, T., 2002, Volcanoes of the World: an Illustrated Catalog of Holocene Volcanoes and their Eruptions: Smithsonian Institution, Global Volcanism Program Digital Information Series, GVP-3, http://www.volcano.si.edu/world/.

Tamura, Y. and, Tatsumi, Y., 2002, Remelting of an Andesitic crust as a possible origin for rhyolitic magma in oceanic arcs: an example from the Izu--Bonin Arc: Journal of Petrology, 43(6), 1029–1047.

How summit calderas collapse on basaltic volcanoes: new insights from the April 2007 caldera collapse of Piton de la Fournaise volcano.

Laurent Michon[1], Nicolas Villeneuve[2], Thibault Catry[1,3], Olivier Merle[1,4]

[1] Laboratoire GéoSciences Réunion, Université de la Réunion, Institut de Physique du Globe de Paris, CNRS, UMR 7154 – Géologie des Systèmes Volcaniques, 15 avenue René Cassin, 97715 Saint Denis, France. laurent.michon@univ-reunion.fr

[2] Institut de Recherche pour le Développement, US 140, BP172, 97492 Sainte-Clotilde cedex, France.

[3] Dipartimento di Scienze della Terra e Geologico Ambientali, University of Bologna, Piazza di Porta S. Donato 1, 40127, Bologna, Italy.

[4] Laboratoire Magmas et Volcans, UMR 6524 CNRS-IRD-Université Blaise Pascal, 5 rue Kessler 63 038 Clermont-Ferrand, France.

Keywords: caldera, GPS, collapse dynamics, Piton de la Fournaise

1. INTRODUCTION

Basaltic volcanoes present summit calderas, whose formation is mostly related to lateral magma migration from the magma reservoir. Observations of basaltic calderas worldwide, and the few recorded collapse events, show common structural characteristics and collapse mechanisms whose origins are still debated. First, caldera collapses are contemporaneous with a periodic seismicity. Second, analyses of the surface deformation have long-revealed that collapses are coeval with the centripetal deflation of the edifice. Third, collapse calderas often show peripheral concentric extensional fractures hundreds of metres from the edge of the caldera rim.

In April 2007, Piton de la Fournaise volcano (PdF) experienced a caldera collapse during its largest historical eruption (Michon et al., 2007). This paper aims at determining the relationship between the concentric fractures and the collapse. It also attempts to better understand the mechanics of the collapse and its consequences in the eruption dynamics. Finally, it addresses the role of the pre-existing structures in the development of new calderas.

2. GEOLOGICAL SETTING

Piton de la Fournaise volcano is one of the world's most active volcanoes. Prior to April 2007, the summit of the active cone was cut by two collapse structures: Bory and Dolomieu, the location of the caldera collapse. On 30th March, a first eruptive fissure opened and remained active during 10 hours (Figure 1a). The magma emission started anew on 2nd April when an eruptive fissure opened at about 600 m asl, 7 km away from the summit. The discharge of magma increased during April 5th contemporaneously with a summit centripetal deflation. The caldera collapse occurred between April 5th and 7th. During this period, both the seismic signal and the summit displacements were organised in cycles characterised by an inward deflation coeval to the increase of the seismic signal, and ended with a sharp outward uplift of the caldera rim contemporaneous with the sudden decrease of the seismicity. The eruption continued at a low level until the 1st May 2007.

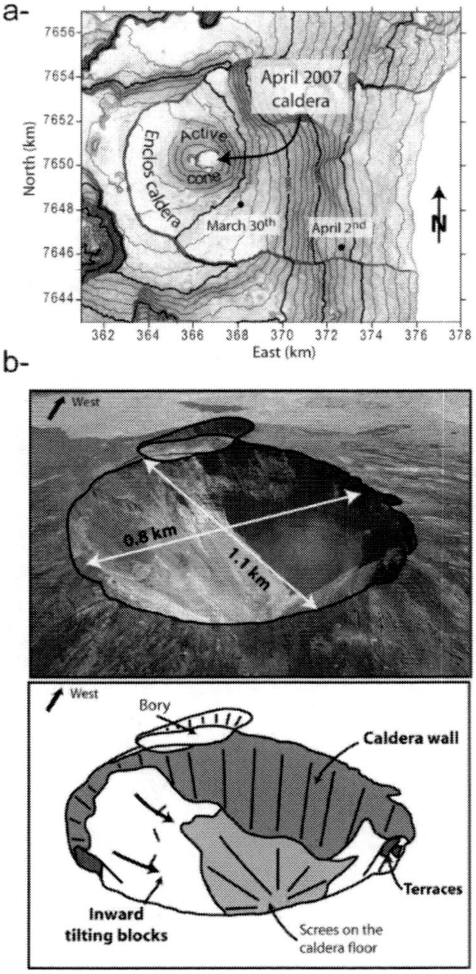

Figure 1: a- Location of the April 2007 caldera and of the eruptive fissures related to the April 2007 eruption. b- Geometry of the caldera.

3. SUMMIT DEFORMATION RELATED TO THE APRIL 2007 ERUPTION

The summit deformation related to the caldera collapse was investigated by combining structural and precise GPS data. GPS campaigns in March and May 2007, *i.e.* prior and after the collapse, reveal a strong inward deflation of the caldera rim, with displacement values ranging between 10 cm to 253 cm. Displacement vectors suggest the occurrence of an E-W elongated deflation source located in the northern part of the caldera. The vertical geometry of the deflation source is further constrained by the ratio between horizontal and vertical displacements, which suggests a vertically elongated structure. Moreover, the sharp decrease of the displacement values from the caldera edge indicates that the first tens of meters of the caldera rim, experienced radial extension stresses. This zone of contrasted displacement is correlated with the development of a new network of concentric fractures.

The new caldera is characterised by a maximum depth of 320-340 m. Despite an intense vertical deformation, the caldera presents the same contour than that of the pre-existing Dolomieu. The first collapses occurred in the northern part of the structure, where the depression was bounded by sub-vertical scarps. This zone is superimposed to the deflation source inferred from GPS data. The caldera subsequently widen by the inward collapse of annular plateaus. This twofold evolution explains the two different geometries of the caldera walls, i.e. sub-vertical, 70-80°, or steep, 40-50° (Figure 1b). Note that remaining polished planes were observed on the sub-vertical walls, suggesting that they correspond to caldera ring faults.

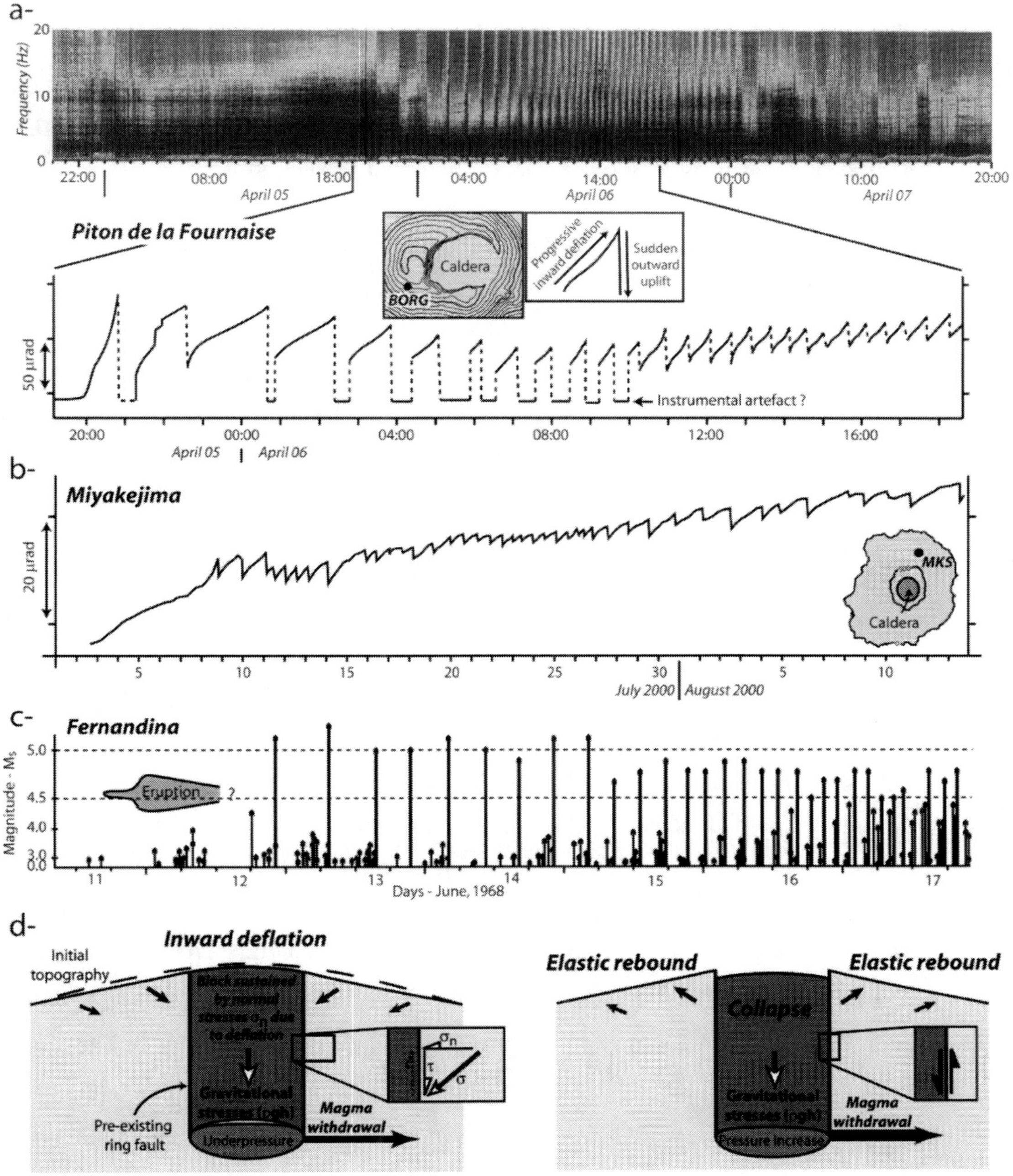

Figure 2: a-Seismic signal and radial deformation measured by tiltmeter (after Staudacher et al., submitted) at Piton de la Fournaise during the April 2007 caldera collapse. BORG: location of both stations. b- Deformation data (radial tiltmeter) recorded during the collapse at Miakejima (after Ukawa et al., 2000). MKS: location of the station. c- Seismicity related with the caldera collapse at Fernandina (after Filson et al., 1973). d- Conceptual models explaining the pulsating dynamics of caldera collapse.

4. DYNAMICS OF CALDERA COLLAPSE

Comparison of the caldera collapses at PdF, 2007, Miyakejima, 2000, and Fernandina, 1968, reveals striking similarities. Deformation and seismic data show that the caldera collapses have been characterized by a cyclic dynamics (Figures 2a, 2b and 2c). Each cycle corresponds to a progressive inward deflation coeval with an increasing seismicity, ended by a sudden radial outward motion when the collapse occurs. Hence, the similar evolution at Fernandina, Miyakejima and PdF strongly suggests that each caldera collapse occurred in a similar way. Combining the different characteristics of these caldera collapses, we propose a unifying mechanism of caldera formation in basaltic setting.

First of all, the deformation of the summit of basaltic volcanoes starts when large lateral eruptions or magma intrusions occur. The pressure decrease into the magma chamber caused by the magma withdrawal entails, above all, the centripetal subsidence of the edifice and subsequently a stress increase within the edifice (Figure 2d). The centripetal deflation prevents the downward motion of the rock column due to gravitational stress by increasing the normal stress along the pre-existing ring faults. The collapse of the rock column occurs when the gravitational stress exceeds the shear strength of the ring faults. The immediate, short outward deformation of the caldera rim after the collapse corresponds to the elastic response of the edifice, when the downward stress is sporadically released during the motion of the rock column. The collapse stops when the pressure increase into the magma chamber sustains the rock column anew. The ongoing subsidence, which directly results from the magma withdrawal, promotes a new stress increase and prevents the collapse of the rock column. The collapse occurs anew when the shear strength is overcome by the gravitational stress exerted on the rock column. The pulsating dynamics of caldera collapse consequently results from a competition between deflation, which prevents the collapse, and gravity exerted on the block, which makes it possible despite deflation.

REFERENCES

Filson, J., Simkin, T., Leu, L.-K., 1973, Seimicity of a caldera collapse: Galapagos Islands 1968. Journal of Geophysical Research, 78(35), 8591-8622.

Michon, L., Staudacher, Th., Ferrazzini, V., Bachèlery, P., Marti, J., 2007, April 2007 collapse of Piton de la Fournaise: a new example of caldera formation. Geophysical Research Letters, 34(21), L21301, doi:10.1029/2007GL031248.

Staudacher, Th., Ferrazzini, V., Peltier, A., Kowalski, Ph., Boissier, P., Catherine, Ph., Lauret, F., Massin, F., The April 2007 eruption and the Dolomieu crater collapse, two major events at Piton de la Fournaise (La Réunion Island, Indian Ocean). Journal of Volcanological and Geothermal Research, submitted.

Ukawa, M., Fujita, E., Yamamoto, E., Okada, Y., Kikuchi, M., 2000, The 2000 Miyakejima eruption: crustal deformation and earthquakes observed by the NIED Miyakejima observation network. Earth Planets Space, 52, xix-xxvi.

IOP Conf. Series: Earth and Environmental Science **3** (2008) 012026 doi:10.1088/1755-1307/3/1/012026

Tectono-volcanic control of fissure type vents for the 28 Ma Panalillo Ignimbrite in the Villa de Reyes graben, San Luis Potosí, México.

Tristán-González, Margarito [1-2], Labarthe-Hernández, Guillermo [1], Aguirre-Díaz, Gerardo J.[2], Aguillón-Robles, Alfredo [1]

[1]Instituto de Geología/DES Ingeniería, UASLP, Av. Dr. Manuel Nava #5, Zona Universitaria, C.P. 78240, San Luis Potosí, S.L.P., MÉXICO. E-mail: **mtristan@uaslp.mx**

[2]Centro de Geociencias, Universidad Nacional Autónoma de México, Campus Juriquilla, Querétaro, 76230 MEXICO. E-mail ger@geociencias.unam.mx

keywords: caldera, structure, collpase, tectonics

1. Introduction.

In the SE portion of the Sierra Madre Occidental, near Sans Luis Potosí, central Mexico, there are several grabens and half-grabens developed during the Basin and Range extensional event. (Stewart 1978, 1998; Aguirre-Díaz et al., 2008). One of these is the Villa de Reyes graben (Fig. 1), which is 200x15 km and with two main orientations, a southern sector oriented NE-SW and a northern sector oriented NNW. This work describes the geology and structure of this graben and its relationship with widespread ignimbrite volcanism.

2. Geology and Structure

Along the main fault scarps of the Villa de Reyes graben (VRG) it is exposed a mid-Tertiary volcanic sequence (32-21 Ma) corresponding to the Sierra Madre Occidental province. According to Labarthe et al. (1982), this sequence overlies marine Mesozoic sedimentary rocks that can be observed at the base of the graben's fault scarps. The volcanic sequence (Fig. 2) includes 32.5 Ma rhyolitic ignimbrites (Santa María ignimbrite), 31.5 Ma dacitic lava domes (Portezuelo dacite), 31 Ma rhyolitic lava domes (San Miguelito rhyolite), 29 Ma ignimbrites (Cantera and San José ignimbrites), and 29 Ma rhyolitic domes (El Zapote rhyolite). Small undated gabbroic plutons are observed cutting the sequence.

In another small graben, Bledos (Fig. 1), oblique to VRG and oriented NW, the exposed sequence includes the Panalillo ignimbrite (Labarthe et al., 1982), a widespread marker unit in this region that is dated at 28 Ma (Torres-Hernández *et al.*, 2008). In the Bledos area also crops out the 28 Ma La Placa basalt. This basaltic-rhyolitic volcanism at around 28 Ma has been interpreted as bi-modal volcanic event (Torres-Aguilera, 2005). Another oblique graben to VRG, also NW, is Las Enramadas graben (Fig. 1). Las Enramadas depression was filled with the nonwelded member of the Panalillo ignimbrite. Pyroclastic feeder dykes along the Las Enramadas graben master faults were the vents for the Panalillo ignimbrite in this locality (Aguirre-Díaz et al., 2008). However, the same fissural style has been observed for the nonwelded Panalillo ignimbrite at other sites in the Río Santa María region (Torres-Hernández et al., 2006; Aguirre-Díaz et al., 2008).

On the basis of water well logs in the VRG, we know that the Mesozoic marine basement is 300 to 574 m deep. At the fault scarps (Figs. 3 and 4), basement exposures occur at the base of them and can reach up to 100 m in elevation. Thus, vertical offset for the VRG faulting is in the order of 600 m. The age of faulting is constrained by the 28 Ma Panalillo ignimbrite that filled portions of the VRG and Enramadas grabens; therefore, faulting initiated before 28 Ma. However, in the area there are tilted plateaus made of Panalillo ignimbrite, indicating that faulting has continued after eruption of Panalillo ignimbrite.

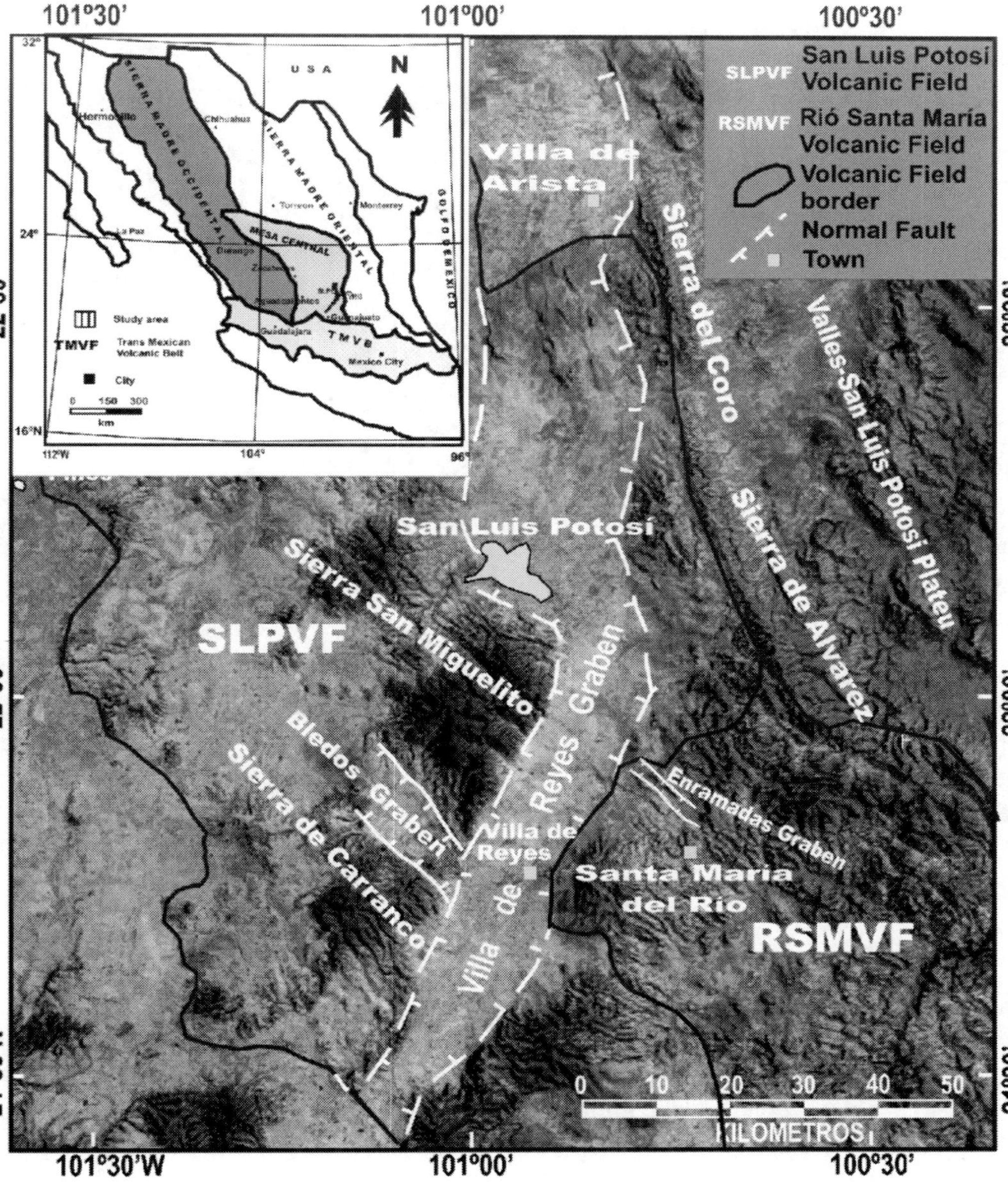

Figure 1. Regional index map (inset) of the Villa de Reyes graben, and geologic provinces Sierra Madre Occidental, Sierra Madre Oriental, Mesa Central, and Transmexican Volcanic Belt (TMVB). Also shown are the locations of Bledos and Enramadas grabens, as well as other features mentioned in text.

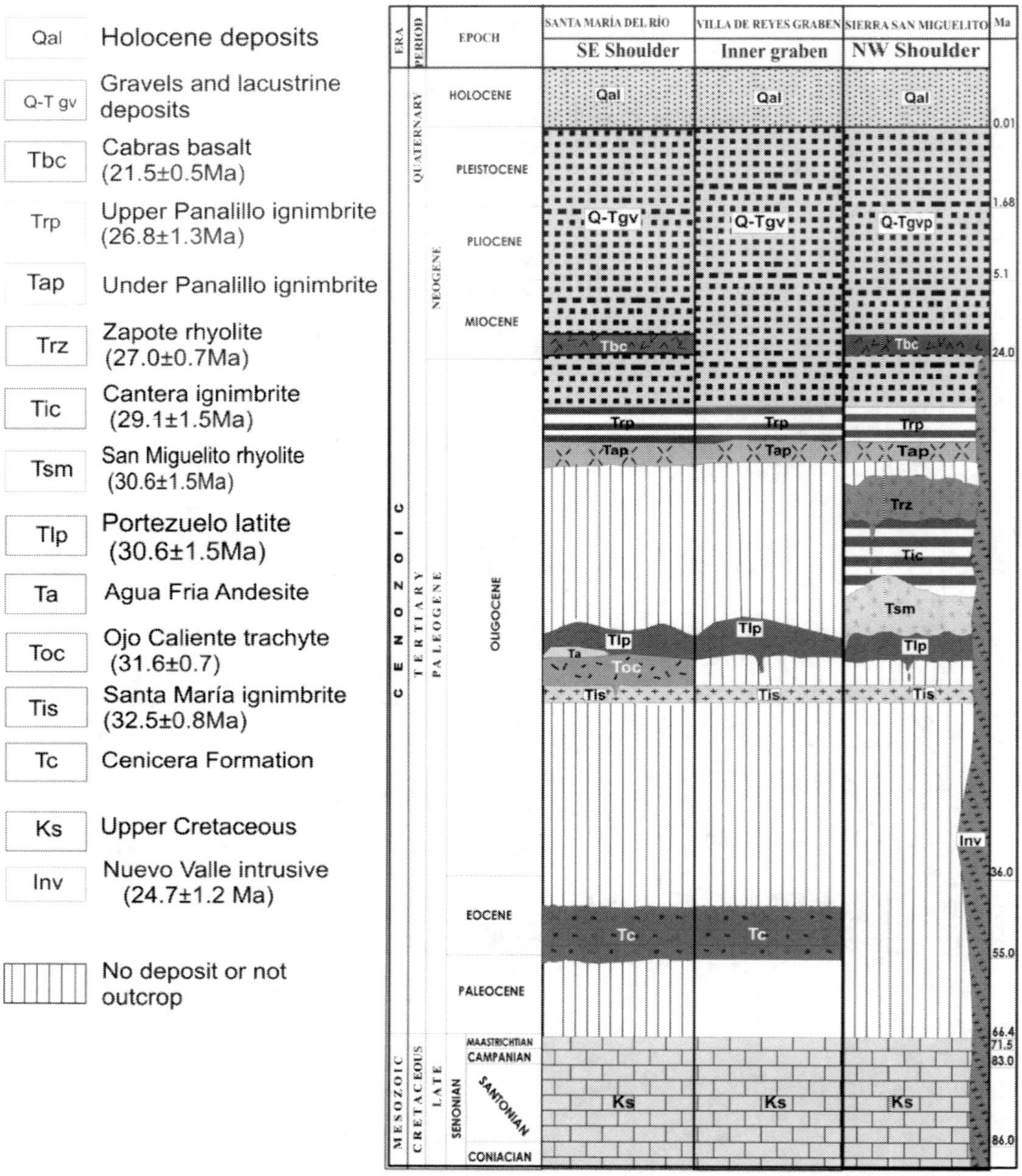

Figure 2. General stratigraphy of the Villa de Reyes graben and surrounding area.

As mentioned above, the nonwelded member of Panalillo ignimbrite was erupted from fissural vents along the faults of the NW-SE oriented grabens that are transversal to the VRG, such as Las Enramadas or at Bledos grabens. There is no evidence that the same happens with the NNW to NE faults of VRG. In the NW Las Enramadas and Bledos grabens there are both pyroclastic dykes made of the same material as the nonwelded Panalillo ignimbrite and co-ignimbrite lithic breccias near these fissural vents (Aguirre-Díaz and Labarthe-Hernández, 2003).

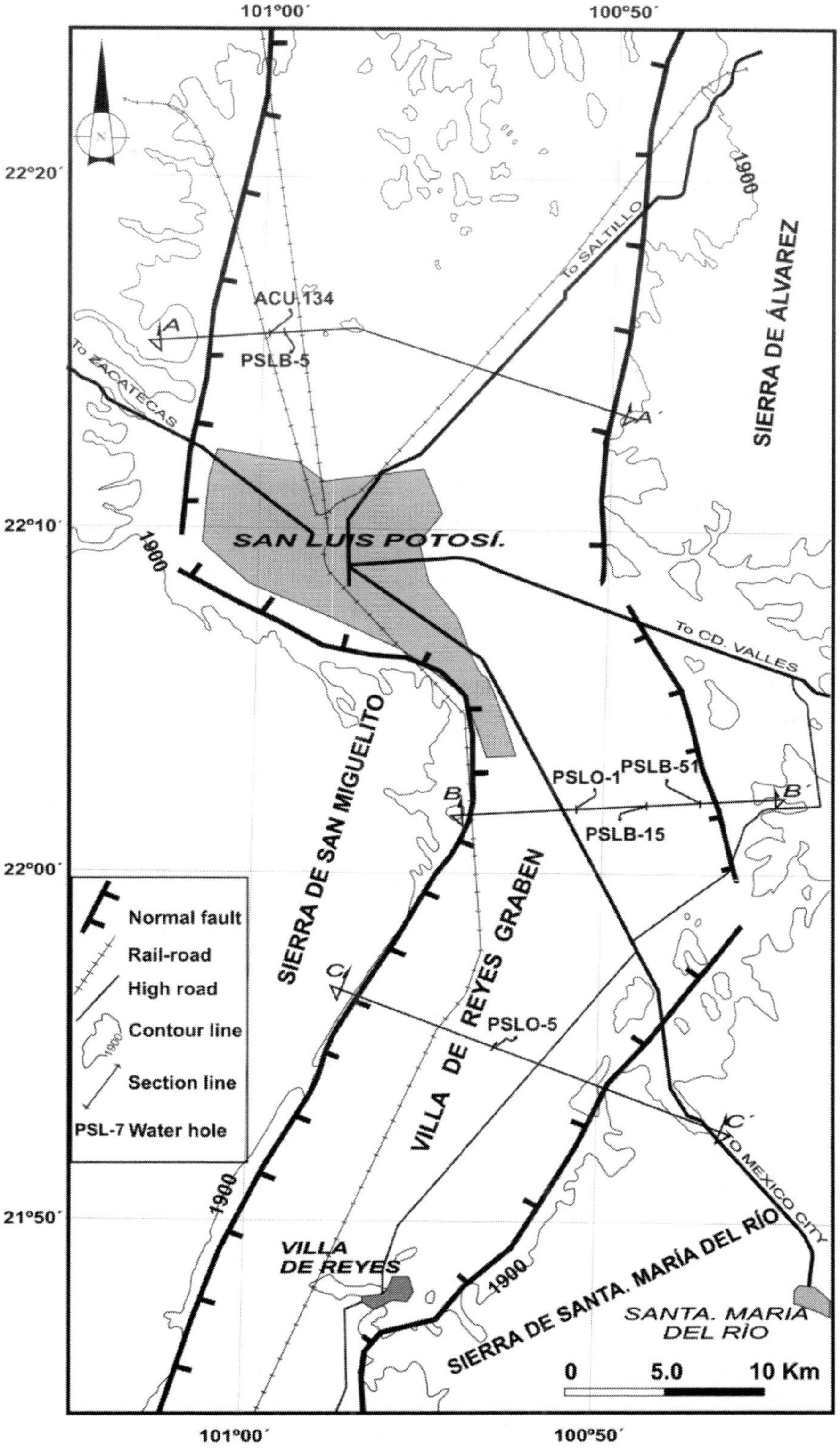

Figure 3. Outline of the general structure of the Villa de Reyes graben and location of geologic cross sections shown in Fig. 4.

The VRG originated during the continental scale extension event known as Basin and Range, which has been related to the mid-Tertiary ignimbrite volcanism that formed the Sierra Madre Occidental province (Aguirre-Díaz and Labarthe-Hernández, 2003). For the area of VRG it is confirmed the overlapping in space and time of both provinces at the time range of 28-26 Ma (Labarthe-Hernández *et al.*, 1982, Tristán-González, 1986; Torres-Hernández et al., 2008). The volcano-tectonic events at VRG are summarized as follows:

1. A regional NNE fault system developed before 32 Ma, which is the age of the oldest rocks affected by this faulting. It is better observed to the North-Northeast of the City of San Luis Potosí.
2. This pre-32 Ma faulting controlled the emplacement of 31.5 Ma dacitic domes (Portezuelos dacite).
3. NE faulting at 28 Ma that displaced the 31.5 Ma dacitic domes and formed the VRG, as well as the oblique grabens of Bledos and Enramadas oriented NW.
4. Emplacement of Panalillo ignimbrite at 28 Ma filling the VRG and erupting from fissures related to the oblique grabens, and eruption of Placa basalt apparently also from fault-controlled vents.
5. Filling of VRG by continental clastic sediments.

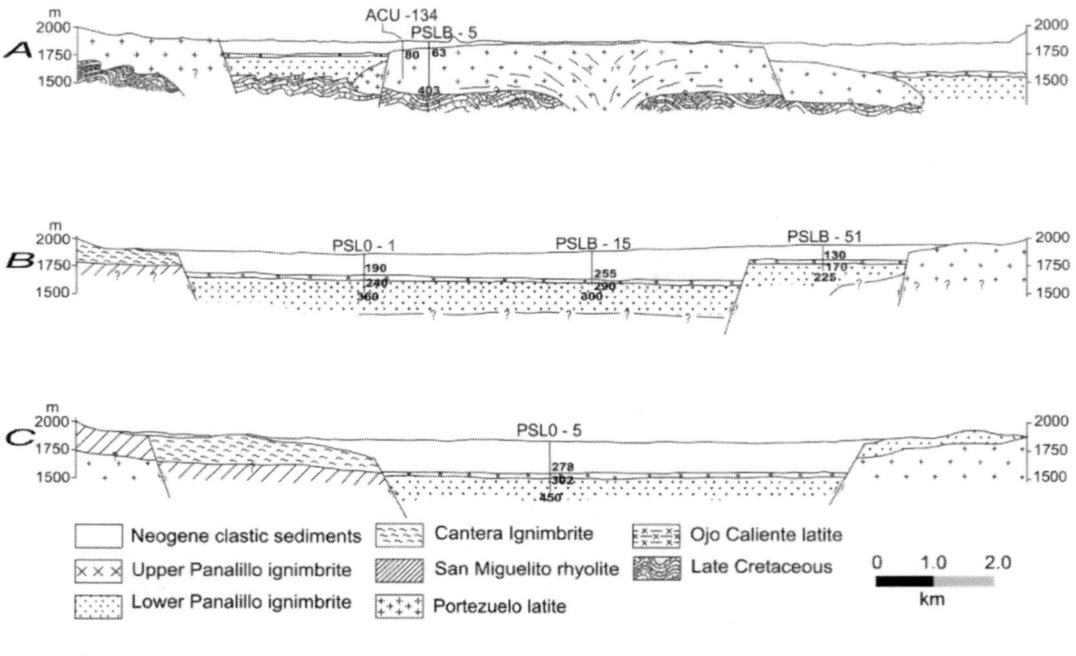

Figure 4 Representative geologic cross sections of the Villa de Reyes graben (see Fig. 3 for location).

IOP Conf. Series: Earth and Environmental Science **3** (2008) 012026 doi:10.1088/1755-1307/3/1/012026

References

Aguirre-Díaz, G.J., and Labarthe-Hernández, G., 2003, Fissure ignimbrites: Fissure-source origin for voluminous ignimbrites of the Sierra Madre Occidental and its relationship with Basin and Range faulting. Geology, v. 31, p. 773-776.

Aguirre-Díaz, G.J., Labarthe-Hernández, G., Tristán-González, M., Nieto-Obregón, J., Gutiérrez-Palomares, I., 2008, Ignimbrite Flare-up and graben-calderas of the Sierra Madre Occidental, Mexico: in J. Martí and J. Gottsmann, eds., "Caldera Volcanism: Analysis, Modelling and Response", Developments in Volcanology 10, Elsevier, Amsterdam, 492 p., (143-180 pp.).

Labarthe-Hernández, G., Tristán-González, M., 1978, Cartografía Geológica Hoja San Luis Potosí: Universidad Autónoma de San Luis Potosí, Instituto de Geología y Metalurgia, Folleto Técnico, 59, 41 p.

Labarthe-Hernández, G., Tristán-González, M., Aranda-Gómez, J.J., 1982, Revisión estratigráfica del Cenozoico de la parte central del Estado de San Luis Potosí: Universidad Autónoma de San Luis Potosí, Instituto de Geologia y Metalurgia, Folleto Técnico, 85, 208 p.

Labarthe-Hernández, G., Tristán-González, M., Aguillón-Robles, A., 1984, Cartografía geológica 1:50,000 Hoja Salitrera, S.L.P: Universidad Autónoma de San Luis Potosí., Instituto de Geología y Metalurgia, Folleto Técnico no. 94, 85 p.

Stewart, J.H., 1978, Basin and Range structure in western North America a review: Geological Society of America Memoir, 152, 1-30.

Stewart, J.H., 1998, Regional characteristics, tilt domains, and extensional history of the late Cenozoic Basin and Range province, western North America: Geological Society of America Special Paper, 323, 47-74.

Torres-Aguilera, J.M., 2005, Caracterización petrográfica y geoquímica del vulcanismo bimodal en el semigraben de Bledos, en el Campo Volcánico de San Luis Potosí: San Luis Potosí, Méx., Universidad Autónoma de San Luis Potosí, [tesis de maestría].

Torres-Hernández, J.R., Labarthe-Hernández, G., Aguillón-Robles, A., Gómez-Anguiano, M., Mata-Segura, J.L., 2006, The pyroclastic dikes of the Tertiary San Luis Potosí volcanic field: Implications on the emplacement of Panalillo ignimbrite: Geofisica Internacional, 45, 243-253.

Torres-Hernández, J.R., Aguillón-Robles, A., Rodríguez-Ríos, R., Bellon, H., Labarthe-Hernández, G., 2008, Un patrón semianular en el emplazamiento de un conjunto de domos riolíticos en el Campo Volcánico de San Luis Potosí: Revista Mexicana de Ciencias Geológicas (in review).

Tristán-González, M., 1986, Estratigrafía y tectónica del Graben de Villa de Reyes, en los estados de San Luis Potosí y Guanajuato, México: Universidad Autónoma de San Luis Potosí, Instituto de Geología, Folleto Técnico, 107, 91 p.

How piecemeal is your caldera? - going beyond modelling to investigate the structural evolution of explosive caldera volcanoes

Chris Willcox[1], Mike Branney[1], Gerardo Carrasco-Núñez[2]

[1]University of Leicester, UK, LE1 7RH
Email: cpw14@le.ac.uk

[2]Centro de Geociencias
Campus Juraquilla, UNAM, Apdo. Postal 1-742
Querétaro, Qro. 76001, México

Keywords: Caldera, Subsidence, Structure, Modelling, Los Humeros

1. Beyond modelling

Modelling is a useful approach in the quest to understand caldera structural evolution. A recent profusion of analogue models has highlighted possible alternative subsidence architectures and parameters that may influence them. However, despite these studies, relatively little is known about the internal structure of explosive caldera volcanoes and how they evolve with time. So how can we now take understanding beyond modelling? How can modellers test their predictions and apply their results, given that the internal structures of many young calderas are concealed?

2. Caldera faults revealed

One approach has been to map and analyse the structures of exhumed ancient caldera floors, where analysis of the field relations can be advantageous in reconstructing the progress of the deformation, for example during a single protracted eruption. Case-studies of this type have revealed considerable diversity: some caldera floors are relatively coherent (e.g. Hwasan, Korea), some have restricted piecemeal zones (e.g. Glencoe: Kokelaar and Moore, 2006), and others have extensive piecemeal zones that fragmented, foundered and tilted in various orientations during a single explosive eruption (e.g. Scafell caldera: Branney and Kokelaar, 1994). In some cases, the structure also can increase further in complexity with overprinting and reactivation during successive super-imposed collapse events. Many exhumed calderas show marginal extension zones similar to those in analogue models (Branney, 1995). Many caldera floors show downsag (inward tilting at margins) much steeper than in analogue models; yet other calderas show negligible downsag. As yet, we cannot account for these differences, but we predict that factors like caldera floor thickness (base of the fill to the magma chamber roof), magma chamber morphology (single cavity or dyke-sill network?), strain-rates of subsidence (related to eruptive mass flux), the strength and degree of anisotropy of the country rock (e.g. layered or massive, sedimentary or crystalline, faulted or pristine), and the tectonic setting (e.g. rates of transtension or extension) are important. A major long-term objective is to understand the causes of diversity sufficiently that we can 'read' calderas: for example to interpret the tectonism, for informing geothermal and mineral prospecting, and for making sensible predictions about likely future eruptive behaviour. Just as with real calderas, analogue models show development of varying degrees of piecemeal internal structure (Figure. 1).

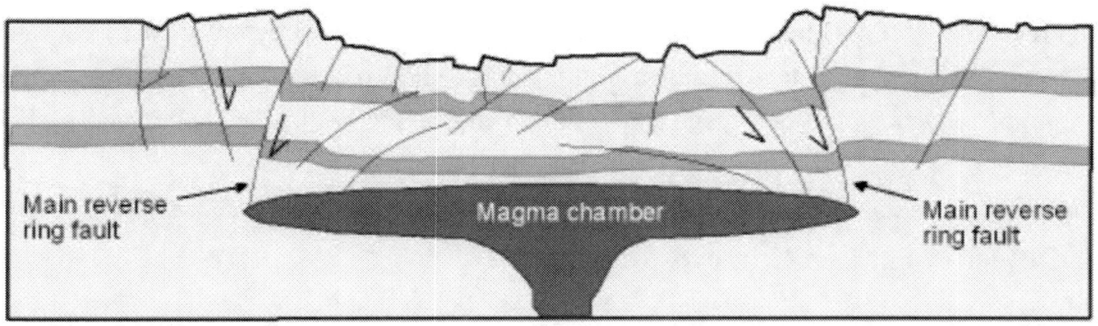

Figure 1: Summary sketch of analogue models of a caldera with well-developed piecemeal structure formed during a single subsidence event in the absence of far-field (regional) stress (adapted from Holohan et al. 2005).

In the last decade the number of analogue simulations has far exceeded the number of detailed case-studies of well-exposed exhumed caldera floors. This imbalance leaves an urgent need for more, well-constrained case-studies. The problem is not limited exposure or ambiguity in interpretation: rigorous analysis of well-exposed field relations can yield valuable and often unexpected structural data (e.g. piecemeal block-faulting caldera subsidence during eruption of the Pagosa Peak Dacite: Bachmann et al. 2000; subsidence on rectilinear rather than arcuate or ring faults at Glencoe: Kokelaar and Moore 2006). The problem is the large investment of time and expertise required to map and analyse an exhumed caldera in sufficient detail to constrain the structural evolution and syn-subsidence tectonism; few volcanologists have time to explore exhumed volcanic terrains that have promise to reveal caldera structures (e.g. the Proterozoic of NW Canada: Hildebrand, 1984).

3. Modern calderas

Modern calderas are generally poor candidates at which to investigate caldera structural evolution. Most are not exhumed, and most faults are concealed by intra-caldera ignimbrites, lakes and/or lavas, rendering meaningful comparisons to models problematic. So how can one further advance the interpretation of the structural evolution of such calderas? We shall attempt to tackle this at Los Humeros caldera, Central Mexican Volcanic Belt. This volcano was selected because it is non-resurgent, has a well-exposed pyroclastic succession, and has abundant geothermal borehole data from the caldera fill and floor. It is calc-alkaline with two main ignimbrites (<120 km^3), over 20 Plinian – sub-Plinian fall deposits, scoria cones and lavas ranging from high silica rhyolite to basalt, from 0.47 ± 0.04 to 0.02 ± 0.01 Ma (Ferriz and Mahood, 1984, 1987). Los Humeros is thought to be a modern example of a caldera with some piecemeal development: it has been proposed that successive major subsidence events significantly fractured the caldera floor thereby increasing the fracture density, causing a reduction of magma residence times, limiting the degree of differentiation, increasing the eruption frequency, and decreasing the eruption magnitude with time (Ferriz and Mahood, 1984).

4. Objectives at Los Humeros

Our goal is to investigate the extent and timing of caldera floor fracture development at Los Humeros caldera, and to test the Ferriz and Mahood hypothesis that increased fracturing of the caldera affected the residence times of magma, and the frequency and magnitude of eruptions. Our interpretations will draw on improved understanding gleaned from recent analogue models and from field studies of exhumed caldera floors. We shall: (1) investigate the caldera's structure by feature-mapping growth faults in the field and from air photos, by correlating borehole stratigraphy and by locating vent locations from ignimbrites, lavas, spatter ramparts and using isopach and isopleth data from successive fallout deposits; (2) constrain the eruption frequency by dating the fallout deposits and reconstructing the eruption history; and (3) investigate the chemistry of successive eruption-units to determine temporal trends in differentiation. This NERC-funded PhD research project is currently in its first year.

References:

Bachmann, O., Dungan, M.A. and Lipman, P.W., 2000, Voluminous lava-like precursor to a major ash-flow tuff: low-column pyroclastic eruption of the Pagosa Peak Dacite, San Juan volcanic field, Colorado: Journal of Volcanology and Geothermal Research, 98, 153-171

Branney, M.J. and Kokelaar, P., 1994, Volcanotectonic faulting, soft-state deformation and rheomorphism of tuffs during development of a piecemeal caldera, English Lake District: Geological Society of America Bulletin, 106, 507–530

Branney, M.J., 1995, Downsag and extension at calderas: new perspectives on collapse geometries from ice-melt, mining and volcanic subsidence: Bulletin of Volcanology, 57, 303-318.

Ferriz, H. and Mahood, G.A., 1984, Eruption Rates and Compositional Trends at Los Humeros Volcanic Center, Puebla, Mexico: Journal of Geophysical Research, 89, 8511-8524

Ferriz, H. and Mahood, G.A., 1987, Strong Compositional Zonation in a Silicic Magmatic System: Los Humeros, Mexican Neovolcanic Belt: Journal of Petrology, 28, 171-209

Hildebrand, R.S., 1984, Folded cauldrons of the early Proterozoic Labine Group, northwestern Canadian shield: Journal of Geophysical Research, 89, 84296-8440

Holohan, E.P., Troll, V.R., Walter, T.R., Münn, S., McDonnell, S. and Shipton, Z.K., 2005, Elliptical calderas in active tectonic settings: an experimental approach: Journal of Volcanology and Geothermal Research, 144, 119–136

Kokelaar, B.P. and Moore, I.D. 2006, Classical areas of British geology: Glencoe caldera volcano, Scotland. Keyworth, Nottingham: British Geological Survey

Automatic earthquake detection and classification with Continuous Hidden Markov Models: a possible tool for monitoring Las Canãdas caldera in Tenerife

Moritz Beyreuther[1], Roberto Carniel[2,3], Joachim Wassermann[1]

1 DEPARTMENT OF EARTH AND ENVIRONMENTAL SCIENCES (GEOPHYS. OBSERVATORY), LUDWIG MAXIMILIANS UNIVERSITÄT MÜNCHEN, D-80333, GERMANY
2 DIPARTIMENTO DI GEORISORSE E TERRITORIO
UNIVERSITÀ DEGLI STUDI DI UDINE, I-33100, ITALY
3 CURRENTLY AT: INSTITUTO DE GEOFÍSICA, UNIVERSIDAD NACIONAL AUTÓNOMA DE MÉXICO, C. UNIVERSITARIA, 04510 D.F. MEXICO, MEXICO
Corresponding Author, roberto.carniel@uniud.it

Keywords: Hidden Markov Models, low SNR, Tenerife, detection, classification

1 Introduction

The basaltic shield structure of Tenerife (Canary Islands, Spain), built between 12 and 3.3 Ma, is now hosting the 3.5 Ma Las Cañadas volcanic complex (Martí et al., 1994), including a caldera due to a series of collapses (Martí and Gudmundsson, 2000). The most recent structure is the Teide-Pico Viejo volcanic complex (TPV), 200 Ka, that produced both basaltic effusive (e.g. Chinyero, 1909) and phonolitic explosive eruptions (e.g. Montaña Blanca, about 2 Ka).

Several seismic stations have been recently installed to better study anomalous clusters of seismic events below the island, that may be linked to a possible reawakening of the volcano.

An important issue is the possible influence of far (25km<far<75km, i.e. also purely tectonic, non-volcano-tectonic) events on the time features of the seismic noise. Evidence of significant eruptions being triggered in the past by regional earthquakes makes this extremely important for hazard assessment (e.g. Tungurahua, see Tárraga et al., 2007)

On Tenerife, a bi-directional interaction between the time evolution of the continuous seismic noise and the occurrence of (volcano-) tectonic events is postulated, which calls for consistent earthquake catalogues, and in turn for automatic detection and classification methodologies.

2 Research Methods

In this study we adopt speech recognition methodologies, thanks to the intrinsic similarity between seismogram signals and speech signals (Ohrnberger, 2001). In speech recognition Hidden Markov Models (HMMs) are successfully applied for spotting spoken words in continuous audio data. Therefore we propose to use HMMs for spotting earthquakes in continuous seismic data, eliminating the need for any pre-triggering and thus being predestinated for low SNR.

As most classifiers, preprocessing is applied to raw data in order to transform it into some new space where it is easier to distinguish the classes. The selected preprocessed features are then used as input for the calculation of the HMM models. In summary a HMM is a double stochastic process which changes state every time unit t and each time a state is entered and observations are generated (e.g. Young et al., 2002). HMM models are then trained for each class: "near" $<25km$, "far" $25km<far<75km$, "dn" mostly human noise during day time and "nn" mostly natural noise during night time. The resulting double stochastic models are used in the following to calculate the probabilities of unknown continuous seismic input stream.

Being a stochastic process, HMM provide the possibility to add a confidence measure to each classification made, basically evaluating how many times it is more probable to detect an earthquake class than a noise class.

3 Results

HMMs are applied to automatic detection and classification at BDG station (Tenerife) on one month continuous raw seismic data recorded during the period 1. — 30 September 2004 (see Tab. 1).

76% of the earthquakes are correctly detected. IGN catalogued 35 earthquakes in this period, 27 (77%) of them were visible at BDG, 8 were not present — not event as vague signal. On the other hand 7 earthquakes which were not catalogued by IGN were detected and classified.

Table 1

	# of Events	Conf.>2 and SNR>2
missed earthquakes	8	3
correct earthquakes	26	18
	26/34 ≈ 76%	18/21 ≈ 86%
correct noise	323861	597
false alarms	139	3
	139/324000 ≈ 0.04%	3/600 ≈ 0.5%

The low SNR ratio at BDG station suggests to evaluate the performance on a small subset of the data set with high SNR (higher than 2), resulting in 21 earthquakes. We also require the algorithm to give a twice as high probability of detecting an earthquake than of detecting noise, resulting also in 21 detections. 3 earthquakes were missed and there are also 3 false alarms. These 18 highly confident earthquakes contain 86% of the major, i.e. high SNR, earthquakes recorded at BDG station. Therefore 18 of the 26 correct classified earthquakes could be marked as highly confident earthquake detections.

HMM are designed to detect and classify signals of multiple different class identities in one step, easily fast enough to run online. The earthquakes in our dataset split up in 3 far earthquakes and 30 near earthquakes. The algorithm classified all 3 far earthquakes, i.e. 100%, correctly as members of the far class and 27 near earthquakes, i.e. 27/30≈87%, correctly as member of the near class. Certainly the number of tested far earthquakes is too small to draw any conclusions but the high number of correct classifications of near earthquakes seem to be a good foundation for an automatic continuous classification.

4 Conclusions

Seismic catalogues, like IGN, which are based on multi station networks, are truly irreplaceable. However, a significant part of the earthquakes catalogued by IGN are not visible at BDG, due to the low SNR. On the other hand small earthquakes near BDG station are not detected at the IGN seismic network. Therefore the HMM method on a local station can provide better information on earthquakes recorded at that single station. This information about earthquakes measured, together with the analysis of seismic noise, is precise enough to provide more statistical evidence supporting the hypothesis of a relationship between the seismic noise and the occurrence of earthquakes (see Tárraga et al., 2006, Carniel et al., 2008a), and its implications in terms of hazard.

HMM can be also used in an automatic online alert level estimation system, where especially the 86% correct high confident (conf. > 2) earthquakes detections restricted to SNR greater than 2 together should prove to be valuable; the low false alarm rate of 0.5% is a necessary criterion in such a scenario.

References

Carniel R., M. Tárraga, F. Barazza, and A. García. Possible interaction between tectonic events and seismic noise at Las Cañadas Volcanic Caldera, Tenerife, Spain. *Bulletin of Volcanology*, pages — In Press, 2008a. 10.1007/s00445-007-0193-7.

Martí J. and A. Gudmundsson. The las cañadas caldera (tenerife, canary island): an overlapping collapse caldera generated by magma-chamber migration. *Journal of Volcanology and Geothermal Research*, 103: 161—173, 2000. 10.1016/S0377-0273(00)00221-3.

Martí J., J. Mitjavila, and V. Araña. Stratigraphy, structure and geochronology of the las cañadas caldera, (tenerife, canary islands). *Geological Magazine*, 131:715—727, 1994.

Ohrnberger M. *Continous Automatic Classification of Seismic Signals of Volcanic Origin at Mt. Merapi, Java, Indonesia*. PhD thesis, Institut für Geowissenschaften, Universität Postdam, 2001.

Tárraga M., R. Carniel, R. Ortiz, J. M. Marrero, and A. García. On the predictability of volcano-tectonic events by low frequency seismic noise analysis at Teide-Pico Viejo volcanic complex, Canary Islands. *Natural Hazards and Earth System Sciences*, 6: 365—376, 2006.

Tárraga M., R. Carniel, R. Ortiz, A. García, and S. De la Cruz-Reyna. Influence of tectonic events on volcanic activity and implications for pre- eruptive warnings. *Geophysical Research Abstracts 9: 02548, European Geosciences Union, General Assembly 2007, Vienna, Austria, 15 - 20 April 2007*, 2007.

Young S., G. Evermann, T. Hain, D. Kershaw, G. Moore, J. Odell, D. Ollason, D. Povey, V. Valtchev, and P. Woodland. The htk book. Technical report, Cambridge University Engineering Department, 2002. HTK Version 3.2.1.

Shallow sub-surface structure of the Central Volcanic Complex of Tenerife, Canary Islands: Implications for the evolution and the recent reactivation of the Las Cañadas Caldera

J. Gottsmann[1], A.G. Camacho[2], J. Martí[3], L. Wooller[4], J. Fernández[2], A. García[5], H. Rymer[4]

[1] Department of Earth Sciences, University of Bristol, Wills Memorial Building, Queens Road, Bristol BS8 1RJ, United Kingdom
j.gottsmann@bristol.ac.uk
[2] Instituto de Astronomia y Geodesia (CSIC-UCM), Ciudad Universitaria, Pza. de Ciencias, 3, 28040 Madrid, Spain
[3] Institute of Earth Sciences "Jaume Almera", CSIC, Lluís Solé Sabarís s/n, Barcelona 08028, Spain
[4] Department of Earth and Environmental Sciences, The Open University, Walton Hall, Milton Keynes, MK7 6AA, United Kingdom
[5]Department of Volcanology, Museo Nacional de Ciencias Naturales, CSIC, C/ José Gutiérrez Abascal, 2, 28006 Madrid, Spain

Keywords: gravity, inversion, structure, caldera, Las Cañadas, Tenerife

1. Introduction

We present a new local Bouguer anomaly map of the Central Volcanic Complex (CVC) of Tenerife, Spain, constructed from the amalgamation of 323 new high precision gravity measurements with existing gravity data from 361 observations. The new anomaly map images the high-density core of the CVC and the pronounced gravity low centred in the Las Cañadas caldera (LCC) in greater detail than previously available. The LCC resulted from at least three vertical collapses of the Las Cañadas edifice over the past 1 Ma and has undergone reactivation in 2004.

2. New gravity anomaly map

The central volcanic complex is dominated by higher than average values, which produce a local positive anomaly (Fig. 1) of more than 45 mGal centred at the western edge of the Las Cañadas caldera, an area known as Boca Tauce, located at the intersection of two principal structural lineaments oriented at about right angles: the Teno-Santiago rift and the Dorsal ridge.

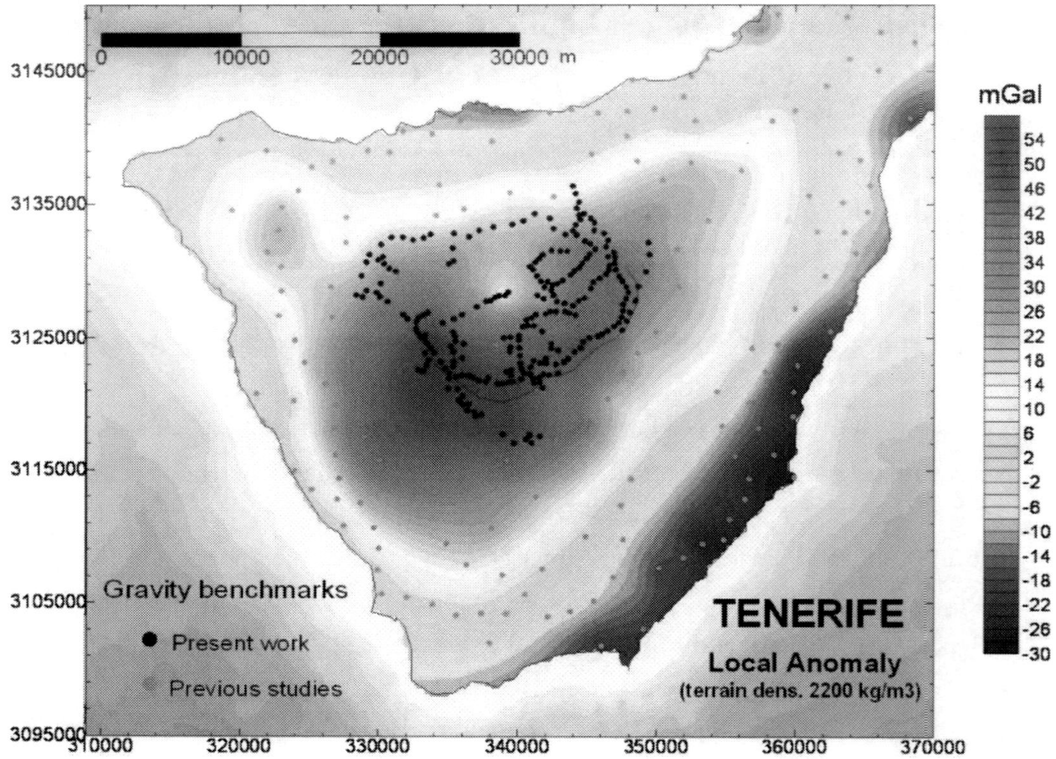

Figure 1: Local gravity anomaly for the central volcanic complex, obtained by subtracting a general linear SE-NW trend of 0.27 mGal/km with azimuth N113°E from the local Bouguer anomaly. This local anomaly data is employed for data inversion. Location of new gravity benchmarks are shown by black points, location of previous benchmarks are shown by gray points.

3. Three dimensional data inversion and construction of sub-surface model

Mathematical construction of a subsurface model from the local anomaly data, employing a 3-D inversion based on "growing" the subsurface density distribution via the aggregation of cells, enables mapping of the shallow structure beneath the complex, giving unprecedented insights into the sub-surface architecture (Figs. 2 and 3).

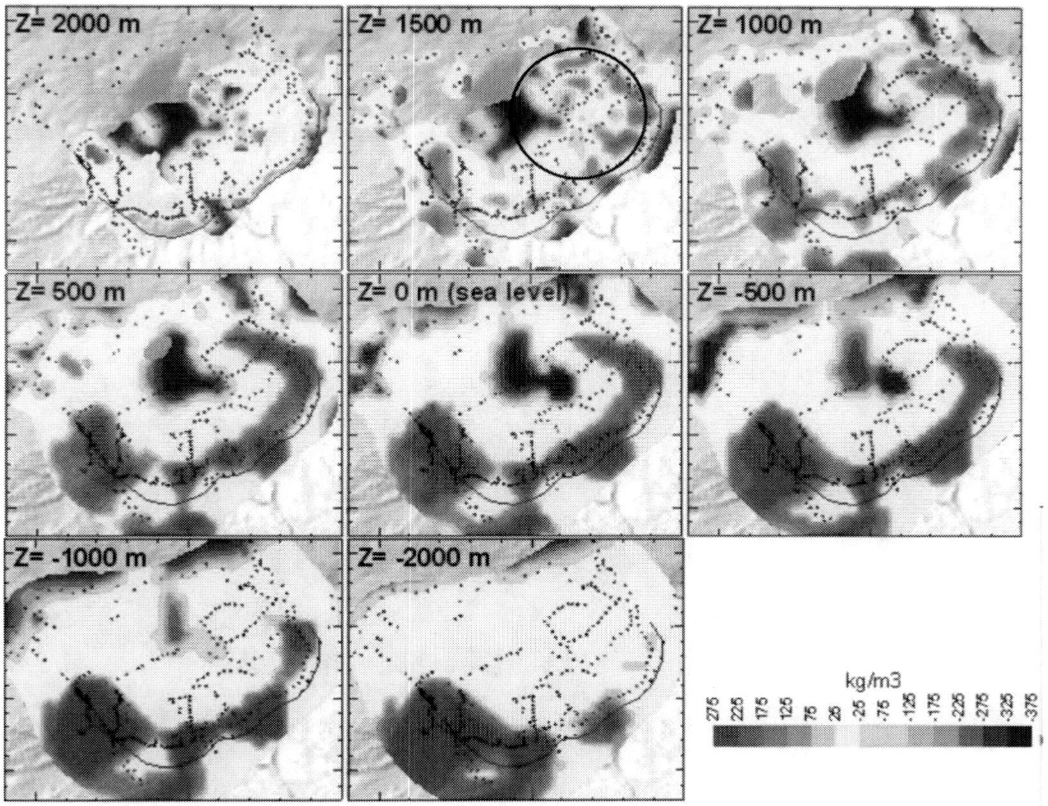

Figure 2: Horizontal sections at selected depths through the 3-D model of density contrasts (kg/m^3) beneath the CVC employing local gravity anomaly data from 323 new gravity measurements. Note the alignment of high density bodies in the eastern part of the caldera (marked by circle), which is interpreted to reflect shallow intrusions along caldera ring faults during the most recent collapse around 179 ka ago.

IOP Conf. Series: Earth and Environmental Science 3 (2008) 012029 doi:10.1088/1755-1307/3/1/012029

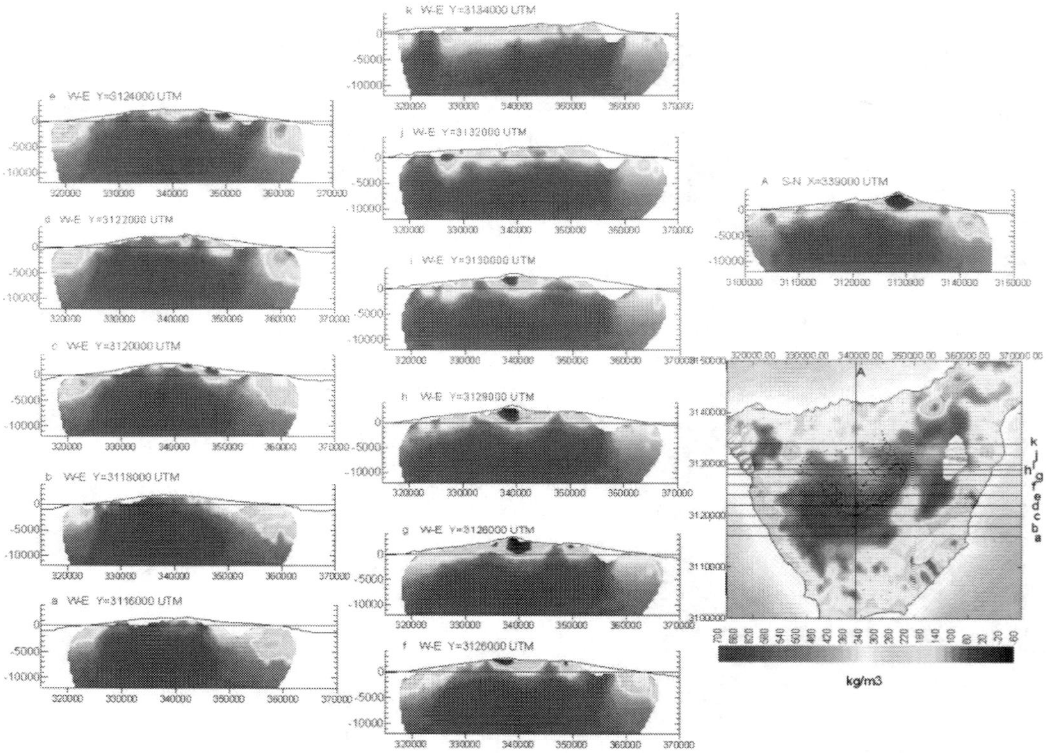

Figure 3: Vertical W-E profiles (a to k) and one S-N profile (A) through the 3D model of density contrasts (kg/m^3) beneath the CVC. Horizontal section indicates trail lines of profiles. Note the pronounced density low beneath the youngest volcanic complex on Tenerife: the Pico Viejo - Pico Teide (PV-PT) complex.

4. Interpretation and summary

Based on a combination of gravimetric and geological data we propose the existence of a central volcanic complex early in the evolution of Tenerife. The phonolitic complex, beneath the centre of the island, most likely formed within an early shield structure. The asymmetric density distribution below sea level is consistent with the removal of significant proportions of the early central edifice in the form of lateral landslides directed to the north in the early history of the Las Cañadas edifice.

We find strong evidence for a vertical collapse origin of the Las Cañadas caldera both from gravimetric data alone as well as from its excellent correlation with results from recent electromagnetic investigations (Coppo et al., 2008). The distribution of small and shallow high-density bodies in the Diego Hernández sector is consistent with intrusions along a ring fault marking the periphery of a funnel-shaped conductive layer interpreted to represent the base of the 179 ka Diego Hernández caldera (Coppo et al., 2008 and Fig. 2).

A horseshoe-shaped alignment of high-density bodies at shallow depth to the north of the Pico Viejo - Pico Teide complex (Fig. 4) is interpreted to represent shallow intrusions into a scar left behind by the Icod landslide, marking the horizontal extent of the slide as well as its head wall. The head wall is nowadays buried beneath eruptive products of the youngest volcanic complex on the island, and thus does not relate in any way to the currently exposed LCC wall. The Icod valley head wall may represent parts of the northern caldera wall, formed by the superposition of three vertical collapses over the past 1 Ma.

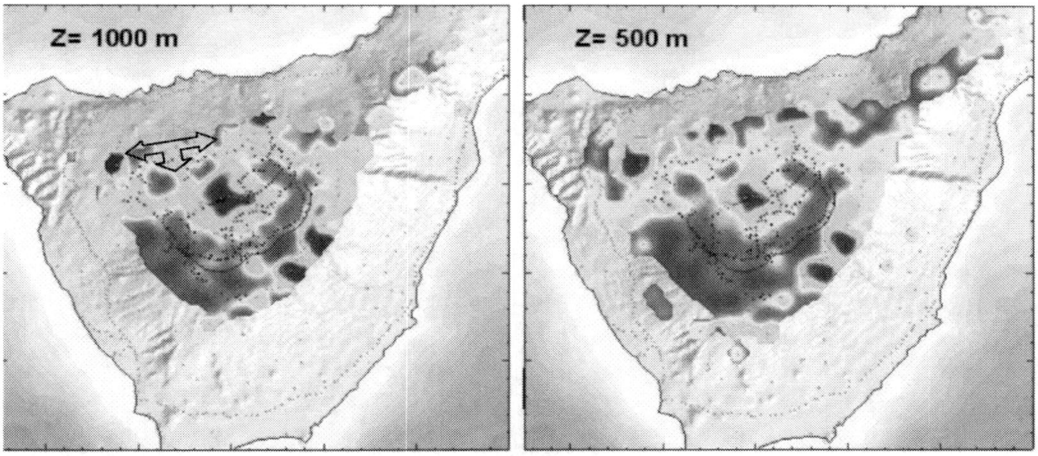

Figure 4: The modelled shallow structure beneath the Central Volcanic Complex for elevations of 1000 and 500 m asl highlights a horseshoe-shaped alignment of high-density bodies, which is interpreted to represent shallow intrusions into a scar left behind by the Icod landslide (marked by arrows).

The interior of the PV-PT complex is characterised by a cylindrically shaped body extending from depths of ca. 3000 m bsl to about sea level. This body of neutral density (Fig. 3) is interpreted to represent the current plumbing system of the complex. Above sea level, low density material dominates the shallow interior of the complex and is thought to result from a combination of evolved eruption products (dominantly phonolitic air-fall and intermediate to evolved vesiculated lavas) with a significant degree of hydrothermal alteration leading to internal corrosion.

The Teno-Santiago rift and the Dorsal ridge have controlled the morphological and volcanic evolution of both the island and the central complex over the past few million years and appear to continue controlling sub-surface dynamics as shown during the recent reactivation of the complex in 2004 (Fig. 5).

IOP Conf. Series: Earth and Environmental Science **3** (2008) 012029 doi:10.1088/1755-1307/3/1/012029

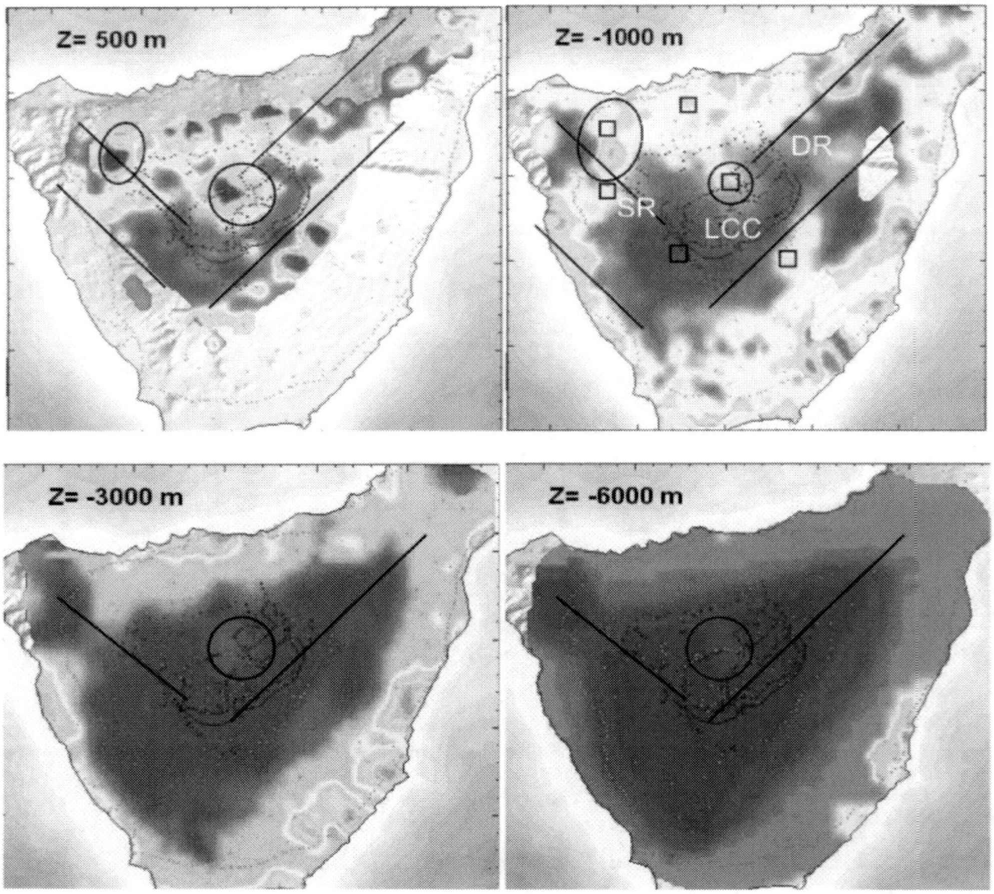

Figure 5: Dominant density anomalies characterising the shallow parts beneath the CVC include i) a NE-SW striking high density ridge forming the southern part of the Las Cañadas edifice with a direct link to the Dorsal Ridge (DR) and a NW-SE striking structure linking the Teno massif to the CVC via the Santiago rift (both marked by lines); ii) two low density bodies (encircled) including (i) the bowl-shaped density low beneath the PV-PT complex, (ii) a NNE-SSW trending low density region located in the Icod valley and crossing the Santiago rift, (iii) a line density lows connecting the Icod (Ic) low to the Tigaiga massif (Tg), and (iv) two density lows to the south of the caldera wall. At depths > 3000 m bsl the high-density core of the CVC resembles the outline of the current island.

Squares in image Z=-1000 m indicate surface projections of positions of ground deformation sources between 2003 and 2006 (Fernández, et al., 2008). It is interesting to note that most source locations are found in areas defined by shallow subsurface negative density anomalies and/or along major structural building blocks of the CVC.

IOP Conf. Series: Earth and Environmental Science **3** (2008) 012029 doi:10.1088/1755-1307/3/1/012029

References:

Coppo, N., Schnegg, P.-A., Heise, W., Falco, P. and Costa, R., 2008. Multiple caldera collapses inferred from the shallow electrical resistivity signature of the Las Canadas caldera, Tenerife, Canary Islands. J. Volcanol. Geotherm. Res., 170 (3): 153-166.

Fernández, J., Samsonov, S., Camacho, A.G., González, P.J., Prieto, J.F., Tiampo, K.F., Rodriguez-Velasco, G., Tunini, L., Willert, V., Charco, M., Mallorquí, J.J., Carrasco, D., 2008. Integration of two line-of-sights classical DInSAR and GPS data to study the 2004-2006 Tenerife volcanic unrest. Geophys. Res. Abstracts, 10, EGU2008-A-10611.

Volcano-tectonic interaction at Soufrière Hills Volcano, Montserrat (W.I.), constrained by dynamic gravity data

Stefanie Hautmann[1*], Joachim Gottsmann[1], Antonio Camacho[2], Nicolas Fournier[3], R. Stephen J. Sparks[1]

[1] Department of Earth Sciences, University of Bristol, Wills Memorial Building, Queens Road, Bristol BS8 1RJ, UK
[2] Instituto de Astronomía y Geodesia (CSIC-UCM), Facultad CC Matemáticas, Universidad Complutense Madrid, 28040 Madrid, Spain
[3] Seismic Research Unit, University of the West Indies, St Augustine, Trinidad, W.I.
[*] corresponding author: stefanie.hautmann@googlemail.com

Keywords: Soufrière Hills Volcano, gravimetry, structural weakness, volcano-tectonic interaction

1. INTRODUCTION

The eruption of Soufrière Hills Volcano (SHV), Montserrat, West Indies, has been ongoing for more than a decade. While geodetic, seismic and geochemical monitoring has been routinely performed, gravimetric surveillance has not been included for most of this activity. Gravity-height time series, however, provide vital information on sub-surface mass and/or density changes in active volcanic areas and are uniquely capable of deducing the nature of causative processes (magmatic or hydrothermal). In June 2006, we began a new joint gravimetric and ground deformation study on Montserrat, with the aim of quantifying mass and/or density changes beneath the island related to the volcanic activity at Soufrière Hills Volcano.

2. SETUP AND DATA COLLECTION

The network was installed and first occupied in June/July 2006 at a time of dome growth of up to 10 m^3/s at SHV. It consists of 10 benchmarks, which cover the accessible part of the island outside the exclusion zone (see Figure 1a). The average distance between neighbouring benchmarks is 2-3 km. The northernmost benchmark, located near Silver Hills is employed as the reference to calculated baselines to all remaining benchmarks. The southernmost benchmarks, those closest to SHV, are located at St. Georges Hill and MVO, at distances of about 3-4 km from the active vent at that time. The network was re-occupied 7 months later, in January/February 2007 during a period of dome growth cessation, which started a few weeks before the survey and was associated with a marked decreased in volcanic activity. Gravity readings during both surveys were carried out at each benchmark manually via optical reading by SH using LaCoste&Romberg gravimeter G-667, while benchmark position and elevation were jointly and simultaneously controlled by Leica500 and Trimble5700 GPS receivers, operating at a frequency of 0.5 Hz.

3. DATA INVERSION AND RESULTS

The GPS data did not reveal significant ground deformation between the two campaigns. Ground deformation was found to be within ±0.02 m, which we interpret as noise for the purpose of the reduction of the gravity data for free-air effects. We therefore propagated a ±5 μGal uncertainty for the free-air effect for our evaluation of uncertainty in the residual gravity data, which we find is on average ±10 μGal. The residual gravity changes between surveys of up to -31 μGal near St. Peters (Figure 1a) are thus reflecting primarily mass changes beneath the central part of the island. Based on the half-width method of the gravity anomaly, the causative source is estimated to be at a maximum depth of about 3 km below sea level assuming a simple spherical geometry of the anomalous body. A Gaussian-integration of the observed mass loss gives $1.03*10^{10}$ kg to $3.55*10^{10}$ kg at a 95% confidence level.

IOP Conf. Series: Earth and Environmental Science 3 (2008) 012030 doi:10.1088/1755-1307/3/1/012030

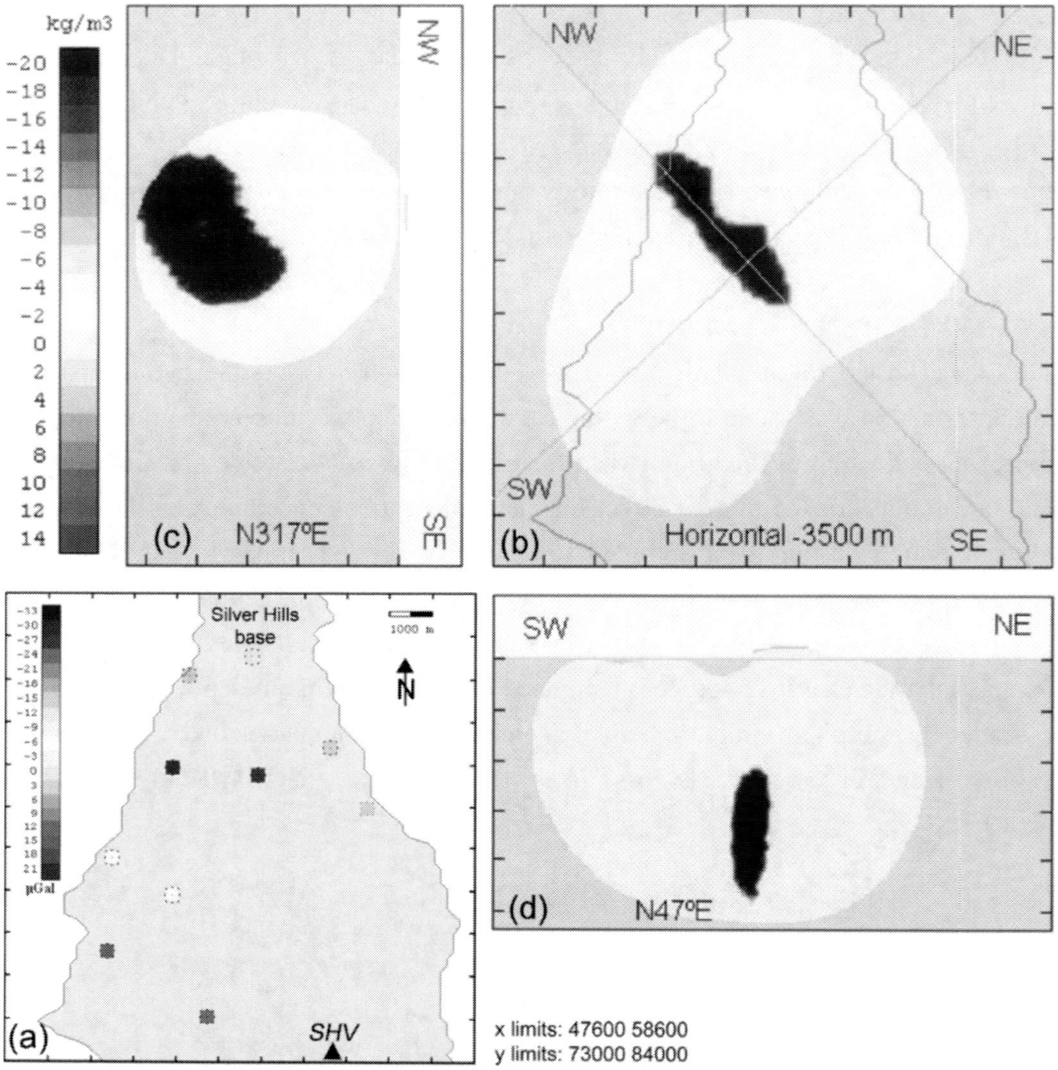

Figure 1: (a) Observed gravity changes at each benchmark, referred to the northernmost benchmark (Silver Hills base). Gravity data were sampled in June/July 2006 and January/February 2007. (b-d) Results of gravity inversion (shown for density contrast -20 to 20 kg/m^3) show a mass reduction in a prolate, elongated, NNW-SSE directed structure, located beneath Central Hills. Top of the structure is at about 2 km depth. Weighted standard deviation of inversion residuals is 4 µGal. Anomalous source is shown in plan view at 3500 m depth (a) and in cross sections, which are orientated NW-SE (b) and NE-SW (c), respectively.

To obtain a model of the subsurface causative source, we performed a non-linear inversion (Camacho *et al.*, 2000, 2002, 2007; Gottsmann *et al.*, 2008) based on a 3-D aggregation of a number of parallelepiped cells, which are filled, in a growth process by means of prescribed positive and/or negative density contrasts. This methodology provides, via an automatic and non-subjective approach, a free 3-D geometry of the anomalous body matching the observed gravity anomaly. Inverting the gravity residuals for given density contrasts, we obtained a set of models, which concur in showing a prolate, NNW-SSE elongated structure located beneath the Centre Hills that underwent mass loss between the two surveys (Figure 1b-d). The top of the source is modelled at about 2 km below sea level. The weighted standard deviation of the inversion residuals is 4 μGal, representing an excellent quality of fit to the data.

4. DISCUSSION

Based on our results, we propose the existence of a previously unrecognised NNW-SSE trending zone of structural weakness (i.e. fault) that is located in shallow depths beneath the Centre Hills of Montserrat. The fault follows the inner volcanic arc and is likely part of the *en echelon* fault zone, which is formed by the Redonda, Bouillante-Montserrat, and Les Saintes fault systems (Feuillet *et al.*, 2002). Surface expressions of the fault are the Soldier Ghaut, a steep and prominent valley in the north of Montserrat and an offset of the shelf-edge NW of the island.

Assigning the observed gravity decrease to magmatic processes in NW Montserrat (dyke emplacement, vesiculation of magma or gas fluctuation in a magma body) would need a magma body, sizing more than 50 Mm^3 in order to fit the recorded signal. The existence of such a magma body appears highly unlikely as no activity is reported from other geophysical records in northern Montserrat. We further exclude shallow fluid migrations along the fault to be the causative source of the recorded changes, as the inferred source depth is located in the water-saturated zone. The changed volcanic activity in between the two microgravity surveys, however, might have resulted in a changed stress field on the island, with a relief of strain leading to a dilation of fractures and an ascent of fluids along the fault, in which they are capped at about 2 km depth (Figure 2). Using a finite element approach we found that a crack opening at depth of up to 12 cm would result in a surface displacement only within the error of GPS measurements. A 12 cm opening of a 5x3 km sized fracture, would lead to a replacement of 1.8*106 m^3 volume solid rock with fluids. Assuming a density difference of 2700 kg/m^3, this conforms to a mass reduction of roughly $5*10^9$ kg, which matches the mass change, derived by integration to within one order of magnitude.

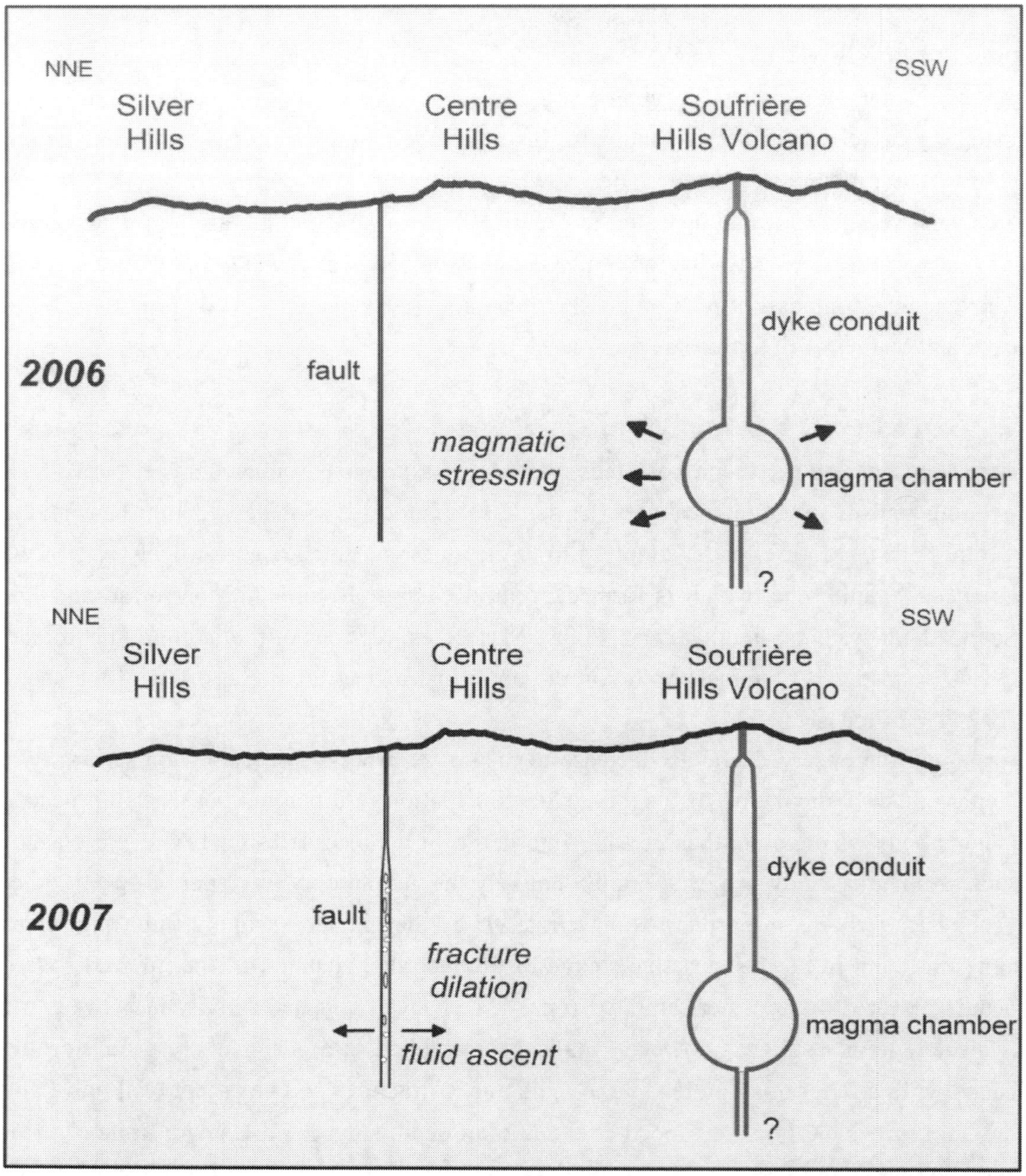

Figure 2: Model of volcano-tectonic interactions as proposed for Montserrat. Fault beneath Centre Hills is locked in times of magmatic stressing (as in 2006), while in times of volcanic quiescence (2007) fractures along the fault dilate and fill up with fluids, which ascend from beneath and are capped at about 2 km depth.

5. CONCLUSIONS

We here propose the existence of a hitherto unrecognized NNW-SSE trending fault zone located beneath the Centre Hills of Montserrat. The fault is linked with SHV volcano via volcano-tectonic interactions. The negative signal recorded in our microgravity survey has been found to be best explained with fracture dilation along the fault zone with a resulting density reduction due to replacement of solid rock material with fluids.

ACKNOWLEDGMENTS

SH acknowledges funding by the Bavarian Research Foundation (BFS DPA-53/05) and the "Arthur von Gwinner-Stiftung" (Max-Planck-Society). JG was supported by a Royal Society University Research Fellowship and NERC grant NE/E007961/1. We are grateful to the staff at MVO for supporting the survey campaigns.

REFERENCES

Camacho, A.G., Montesinos, F.G., Vieira, R., 2000, A 3-D gravity inversion by means of growing bodies: Geophysics 65, 95-101.

Camacho, A.G., Montesinos, F.G., Vieira, R., 2002, A 3-D gravity inversion tool based on exploration of model possibilities: Computer and Geosciences 28, 191-204.

Camacho, A.G., Nunes, J.C., Ortiz, E., França, V., Vieira, R., 2007, Gravimetric determination of an intrusive complex under the island of Faial (Azores). Some methodological improvements: Geophysical Journal International 171, 478-494.

Feuillet, N., Manighetti, I., Tapponnier, P., Jacques, E., 2002, Arc parallel extension and localization of volcanic complexes in Guadeloupe, Lesser Antilles: Journal of Geophysical Research 107(B12), 2331, doi:10.1029/2001JB000308.

Gottsmann, J., Camacho, A.G., Martí, J., Wooller, L., Fernández, J., García, A., Rymer, H., 2008, Shallow structure beneath the Central Volcanic Complex of Tenerife from new gravity data: Implications for its evolution and recent reactivation: Physics of the Earth and Planetary Interiors, doi:10.1016/j.pepi.2008.06.020.

The dispersal of ash during explosive eruptions from central volcanoes and calderas: an underestimated hazard for the Central Mediterranean area

Roberto Sulpizio[1], Benoit Caron[2], Biagio Giaccio[3], Martine Paterne[4], Giuseppe Siani[5], Giovanni Zanchetta[2], Roberto Santacroce[2]
[1] CIRISIVU, c/o Dipartimento Geomineralogico, via Orabona 4, 70125, Bari, Italy; r.sulpizio@geomin.uniba.it
[2] Dipartimento di Scienze della Terra, via S. Maria 53, 56126, Pisa, Italy
[3] Istituto di Geologia Ambientale e Geoingegneria, CNR, Via Bolognola 7, 00138 Rome, Italy
[4] LSCE, Laboratoire Mixte CEA-CNRS-UVSQ, Avenue de la Terrasse 91198 Gif-sur-Yvette Cedex, France
[5] IDES-UMR 8148, Université Paris-XI, 91405 Orsay Cedex, France

Keywords: ash, caldera, hazard

1. INTRODUCTION

The central Mediterranean area comprises some of the most active volcanoes of the northern hemisphere. Some of their names recall myths or events in human history: Somma-Vesuvius, Etna, Stromboli, Vulcano, Ischia and Campi Flegrei. These volcanoes are still active today, and produce both effusive and explosive eruptions. In particular, explosive eruptions can produce and disperse large amount of volcanic ash, which pose a threat to environment, economy and human health over a large part of the Mediterranean area. Volcanic ash (diameter < 2 mm) is the result of intense magmatic or phreatomagmatic fragmentation during explosive volcanic eruptions. After injection into the atmosphere, the ash is dispersed as convective columns and umbrella clouds, which are subject to the combined effects of gravity and wind speed, or are transported close to the ground as pyroclastic density currents (PDCs; e.g. Sulpizio et al., 2008a). Irrespective of the eruptive mechanism or intensity, ash particles usually affect wide areas around volcanic centres, and can cause severe damage and casualties in both proximal and medial areas. The accumulation of ash can induce roof collapses, interruption of lifelines (roads, railways, etc.), closure of airports and damage to communication or electric lines (e.g. Blong, 1984). The injection of ash into the atmosphere can cause damage to aircraft or can impact public health causing, for example, respiratory problems (e.g. Horwell and Baxter, 2006). Ash deposition decreases soil permeability, increases surface runoff, and promotes floods (e.g. Favalli et al., 2006). Ash leachates can result in pollution of water resources (e.g. Stewart et al., 2006), damage to agriculture and forest, effect pasture and livestock health, impinge on aquatic ecosystems and alter the geochemical environment of the seafloor (e.g. Haeckel et al., 2001). Therefore, the impact of fine ash can determine important economic loss for affected communities, and can rapidly deteriorate if no preventive measures were taken. (e.g. Sulpizio et al., 2007).

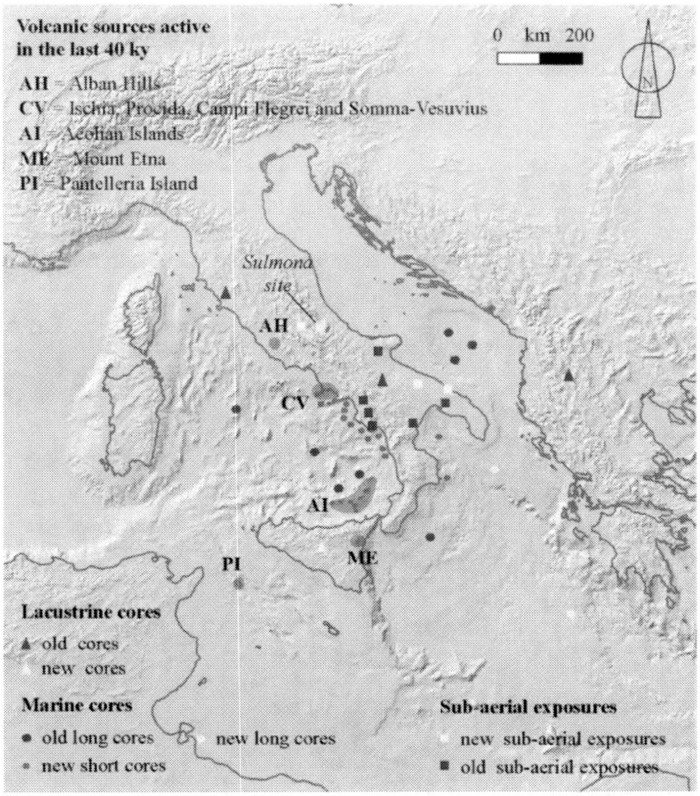

Figure 1: Location map of the cores and sites studied and sampled in the Central Mediterranean area.

All of these impacts and processes have surely affected the central Mediterranean area during past eruptions of Italian volcanoes, as testified by the abundance of ash layers in the stratigraphic successions. Reliable records of proximal, medial and distal ash deposits are in fact available from outcrops, historical data, lacustrine and marine cores. These occurrences, together with the detailed knowledge of the eruptive history of the Italian volcanoes at least for Late Pleistocene and Holocene (e.g. Poli et al., 1987; Rosi and Sbrana, 1987; Santacroce, 1987; Vezzoli, 1988; Di Vito et al., 1999; 2008; Pappalardo et al, 1999; Santacroce and Sbrana, 2003), makes the Central Mediterranean area an exceptional site for detailed studies on ash dispersal.

Despite some recent advances in understanding the impact of fine ash on environment and infrastructure, the dynamic of dispersal of fine ash remains poorly understood, and consideration of the associated hazards have not yet been fully addressed and included in the mitigation plans, with the only exception of the potential impact on air traffic managed by VAACs. This is particularly true for the active volcanoes of southern Italy, whose deposits of past explosive eruptions were dispersed in large parts of central and southern Italy. In many cases these ash were generated during Plinian to sub-Plinian eruptions or final Vulcanian phases of Plinian eruptions, and this oblige to face with ash dispersal over large areas from eruptions with different sizes and intensities.

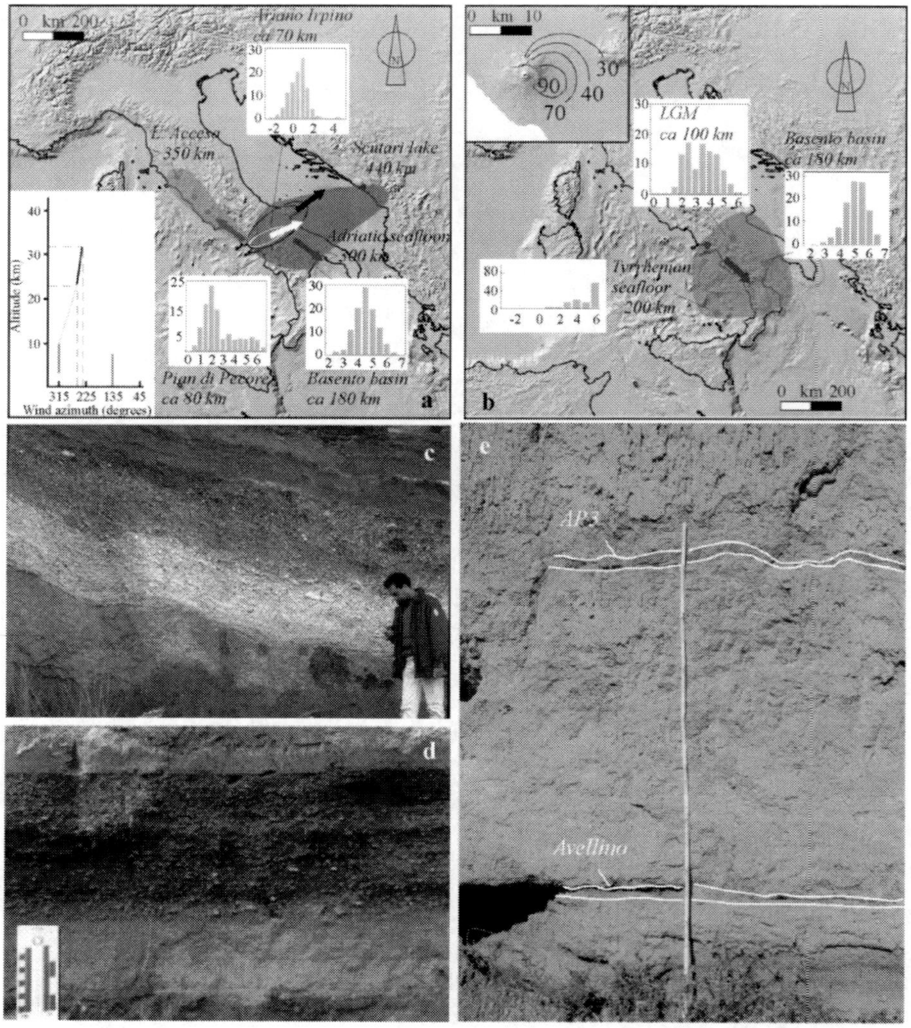

Figure 2: Dispersal, grain size and deposits characteristics of Avellino and AP3 eruptions of Somma-Vesuvius. a) Different dispersal areas of ash from magmatic Plinian (blue shaded area) and final phreatomagmatic (red shaded area) phases of the Avellino eruption. The 1 cm isopachs of white and grey pumice fallout deposits are reported as white and black ellipses, respectively (from Cioni et al., 2000). Arrows indicate the inferred direction of stratospheric and lower atmosphere winds during the different phases of the eruption. Grain size distributions of distal fallout deposits. X axis indicates grain size expressed as funits, while Y-axis indicates % of abundance; b) Dispersal area of AP3 ash deposits (red shaded area): The insert in the upper left corner reports the proximal isopachs (thickness in cm).

Grain size distributions of distal samples from Basento, Lago Grande di Monticchio (LGM) and Tyrrhenian seafloor are reported. X axis indicates grain size expressed as funits, while Y-axis indicates % of abundance (note that the Tyrrhenian seafloor sample is reposrted at 1f interval instead that ½ f;c) proximal fallout deposits of Avellino eruption; d) proximal fallout deposits of AP3 eruption; e) distal Avellino and AP3 ash deposits at Basento site.

As an example, we present and discuss the data of ash dispersal from some explosive eruptions of southern Italy volcanoes, which dispersed centimetric thick ash blankets hundred of kilometres from the source, irrespective of the more limited dispersal of the respective coarse grained fallout and PDC deposits. The collected data also highlight the major role played by lower atmosphere winds in dispersal of ash from weak plumes and ash clouds that accompany PDC emplacement.

2. RESULTS

Distal samples from tens of explosive eruptions occurred in the last 40 ka have been collected in the central Mediterranean area for geochemical and grain size analyses, using already published and unpublished marine and lacustrine cores and sub-aerial exposures (Fig. 1). The collected data allowed the reconstruction of the minimum areas affected by ash deposition during explosive eruptions of different magnitude and intensity from the active volcanoes of southern Italy. As an example, in Figure 2 are reported the dispersal areas of two Holocene eruptions of Somma-Vesuvius volcano. The first is the Avellino Plinian eruption, dated at 3760±70 ^{14}C yr BP (Cioni et al., 2000; Sulpizio et al., 2008b). The Avellino eruption was characterised by three main eruptive phases: opening, magmatic Plinian, and final phreatomagmatic. Magmatic fragmentation driven the eruption dynamics during the opening and Plinian phases, whereas phreatomagmatic fragmentation dominated the final eruptive phase. The first two phases were dominated by fallout deposition, while the final phreatomagmatic phase was dominated by PDC emplacement. The calculated bulk volume of Avellino deposits is at least 1.8 km^3 (1.2 km^3 of fallout deposits and 0.6 km^3 of PDC deposits), and the peak intensities were in the order of 10^7 -10^8 kg/s, which correspond to maximum column heights between 23 and 31 km (Cioni et al., 2000; Sulpizio et al., 2008b). Ash deposits related to the magmatic phases of the eruption were recognised in a wide area (ca. 75,000 km^2) east of Somma-Vesuvius (blue-shaded area in Fig. 2a; Sulpizio et al., 2008a), more than four times the area enclosed by the 1 cm isopach previously hypothesized (Cioni et al., 2000; Fig. 2a). If the shaded area is assumed to represent the 1-cm thickness (reasonable assumption based on thickness of some distal deposits) the distal volume of Avellino fallout deposits is of about 2.6 km^3, which increases the total volume at 3.7 km^3, a value that doubles the previous estimations. Grain size analyses of distal deposits show the prevalence of particles at the transition between coarse and fine ash (main mode at 4.5 f, 44 mm; Fig. 2a), still yet above the size of respirable ash particles. Phreatomagmatic ashes have been recognised over an area of more than 22,000 km^2 to the north west of Somma-Vesuvius (red shaded area in Fig. 2a; Sulpizio et al., 2008a), with thickness of 3 mm up to distances of more than 180 km from the source (Sulpizio et al., 2008a).

The second example here reported is the AP3 eruption (Andronico and Cioni, 2002), dated at 2710±60 ^{14}C yr BP (Rolandi et al., 1998). The AP3 eruption was dominated by magmatic fragmentation, and described deposits comprise only fallout material. The eruption was classified as violent Strombolian to Vulcanian, with calculated peak intensity of 3 x 10^6 kg/s. The proximal fallout deposits (Fig. 2d) are dispersed to the east (Fig. 2b), and have a calculated on land volume of about 1.5 x 10^8 m^3 (Andronico and Cioni, 2002). The AP3 ash deposits were recognised over a wide area (more than 64,000 km^2; Fig. 2b), more than four hundred larger than the area covered by proximal deposits (Fig. 2b).

The dispersal of AP3 ash suggest the combined action of higher and lower atmosphere winds, with the former responsible of the dispersal to the east of the coarse grained proximal fallout deposits (Fig. 2d) and the latter that influenced the dispersal of ash particles to south and south-east (Fig. 2b).

Grain size analyses of AP3 ash deposits (180-200 km from the source) mainly comprise fine ash, with significant fractions of grain size distributions formed by respirable particles.

References

Andronico, D., Cioni, R., 2002. Contrasting styles of Mount Vesuvius activity in the period between the Avellino and Pompeii Plinian eruptions, and some implications for assessment of future hazards. Bull. Volcanol. 64: 372–391.

Blong, R.J., 1984. Volcanic hazards. A Sourcebook on the Effects of Eruptions. Academic Press, Sydney. 424 pp.

Cioni, R., Levi, S., Sulpizio, R., 2000. Apulian Bronze Age pottery as a long-distance indicator of the Avellino Pumice eruption (Vesuvius, Italy). In: Mc Guire W.G., Griffiths D.R., Hancock P.L., Stewart I.S. (eds) The Archaeology of Geological chatastrophes. Geological Society, London, Special Publications, 171, 159-177.

Di Vito, M.A., Isaia, R., Orsi, G., Southon, J., de Vita, S., D'Antonio, M., Pappalardo, L., Piochi, M., 1999. Volcanism and deformation since 12,000 years at the Campi Flegrei caldera (Italy). J. Volcanol. Geotherm. Res., 91, 221-246.

Di Vito, M.A., Sulpizio, R., Zanchetta, G., D'Orazio, M., 2008. The late Pleistocene pyroclastic deposits of the Campanian Plain: new insights into the explosive activity of Neapolitan volcanoes. J. Volcanol. Geotherm. Res. doi:10.1016/j.jvolgeores.2007.11.019.

Favalli, M., Pareschi, M. T., Zanchetta, G., 2006. Simulation of syn-eruptive floods in the circumvesuvian plain (southern Italy). Bull. Volcanol., 68, 349–362.

Haeckel, M., van Beusekom, J., Wiesner, M.G., Konig, I., 2001. The impact of the 1991 Mount Pinatubo tephra fallout on the geochemical environment of the deep-sea sediments in the South China Sea. Earth Plan. Sc. Lett. 193, 151-166.

Horwell, C.J., Baxter, P.J., 2006. The respiratory health hazards of volcanic ash: a review for volcanic risk mitigation. Bull. Volcanol., 69, 1–24.

Pappalardo, L., Civetta, L., D'Antonio, M., Deino, A., Di Vito, M., Orsi, G., Caramente, A., de Vita, S., Isaia, R., Piochi, M., 1999. Chemical and Sr-isotopical evolution of the Phlegrean magmatic system before the Campanian Ignimbrite and the Neapolitan Yellow Tuff eruptions. J. Volcanol. Geotherm. Res., 91, 179-214.

Poli, S., Chiesa, S., Gillot, P.Y., Guichard, F., 1987. Chemistry versus time in the volcanic complex of Ischia (Gulf of Naples, Italy): Evidence of successive magmatic cycles. Contrib. Min. Petrol. 95, 322-335.

Rolandi, G., Petrosino, P., McGeehin, J., 1998. The interplinian activity at Somma-Vesuvius in the last 3,500 years. J. Volcanol. Geotherm. Res. 82, 19–52.

Rosi, M., Sbrana, A. (eds), 1987. Phlegraean Fields. CNR, Quaderni della Ricerca Scientifica 114 (8), 175 pp

Santacroce, R. (ed), 1987. Somma-Vesuvius, CNR Quaderni della Ricerca Scientifica 114(8), Roma, 230 pp.

Santacroce, R., Sbrana, A. (eds), 2003. Carta geologica del Vesuvio. Scala 1:15.000. SELCA editore, Firenze.

Stewart, C., Johnston, D.M., Leonard, G.S., Horwell, C.J., Thordarson, T., Cronin S.J., 2006. Contamination of water supplies by volcanic ashfall: A literature review and simple impact modelling. J. Volcanol. Geotherm. Res, 158, 296-306.

Sulpizio, R., Bonasia, R., Dellino, P., Di Vito, M.A., La Volpe, L., Mele, D., Zanchetta, G., Sadori, L., 2007. Can the Vesuvius' ash threat the city of Rome? Answers from the late quaternary geological record of central-southern Italy. El Chichon volcano: Twenty five years later. A commemorative conference. San Cristobal de las Casas, March 20-23, 2007, abstract volume.

Sulpizio, R., Bonasia, R., Dellino, P., Di Vito, M.A., La Volpe, L., Mele, D., Zanchetta, G., Sadori, L., 2008a. Discriminating the-long distance dispersal of fine ash from sustained columns or near ground ash clouds: the example of the Pomici di Avellino eruption (Somma-Vesuvius, Italy). J. Volcanol. Geotherm. Res., doi:10.1016/j.jvolgeores.2007.11.012

Sulpizio R., Cioni R., Di Vito M.A., Mele D., Bonasia R., Dellino P., La Volpe L., 2008b. The Avellino eruption of Somma-Vesuvius (3.8 ka BP) part I: stratigraphy, chemistry and eruptive mechanisms. Bull. Volcanol. Submitted.

Vezzoli, L, 1988. Island of Ischia. CNR, Quaderni della Ricerca Scientifica 114(10), 230 pp., Roma.

Caldera formation and more, but what comes next? An analysis of volcanic threat of Nisyros island, Greece

A. Winson, H. Kinvig, J. Gottsmann, E. Partington, A. Geyer

Department of Earth Sciences, University of Bristol, United Kingdom

Keywords: hazards, calderas, threat, risk, Nisyros

1. Introduction

The USGS National Volcano Early Warning System (NVEWS) is based on a systematic assessment of the threats posed by active volcanoes coupled with an evaluation of current monitoring capability (Ewart, 2007). Volcanic threat is the combination of hazards (the destructive natural phenomena produced by a volcano) and exposure (people and property at risk from the realisation of hazardous phenomena). The evaluation involves a methodical assessment of 15 hazard and 10 exposure factors including factors such as maximum VEI, eruption recurrence, hydrothermal explosion potential, local aviation exposure and transportation infrastructure. The evaluation is based on a score system, whereby the higher the score, the higher the threat is.

2. Nisyros island

We have applied the catalogue of questions to the volcanic island of Nisyros in Greece, an 8 km-wide stratovolcano located at the eastern end of the Hellenic island arc. The island hosts a 3.8 km-wide caldera and an approx. 0.9 km^2 hydrothermal area with phreatic craters, fumaroles and mudpools. The caldera has been the locus of at least 13 phreatic eruptions in historical times, the most recent in 1888, and is presently affected by considerable hydrothermal activity. A volcano-seismic crisis on Nisyros between 1995 and 1998 marked the reawakening of the volcano and was accompanied by ground uplift and seismicity, which severely damaged infrastructure on the island.

3. Threat analysis

Our NVEWS analysis results in a score of 152 points, taking a cautious and conservative approach to hazard factors due to the lack of absolute ages for the most recent magmatic eruptions. Adverse exposure factors include significant scores relating to aviation and population exposure to volcanic hazards from Nisyros. The score puts Nisyros in the league of volcanoes posing a very high threat. The cut-off score between high threat and very high threat volcanoes in the US is 123 points. US volcanoes with a comparable threat level include Mt. St. Helens, Augustine and the Long Valley caldera. Existing monitoring programs at Nisyros must be described as rudimentary due to the lack of permanent multiparameter surveillance and obtain at best a Level 1 rating according to NVEWS guidelines. The recommendable level of monitoring is level 4 including deployment of instrumentation for continuous seismic and ground deformation surveillance, leaving a monitoring gap of 3 levels. Based on our analysis, we propose that the volcanic threat posed by Nisyros volcano is severely underestimated. It is therefore important to take steps towards a better understanding of the current dynamic behaviour of the volcano, in order to decrease response times in the event of renewed activity and to increase the ability to forecast events, therefore significantly reducing volcanic risk.

Reference:

Ewert. J W., 2007, System for Ranking Relative Threats of U.S. Volcanoes, Natural Hazards Review, 8: 112-124, DOI: 10.1061/(ASCE)1527-6988(2007)8:4112

Institute of Physics
Dirac House, Temple Back
Bristol BS1 6BE UK

CURRAN ASSOCIATES INC
proceedings.com